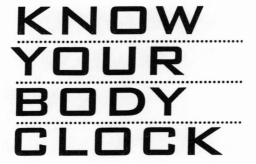

KNOW YOUR BODY CLOCK

KNOW YOUR BODY CLOCK

DISCOVER YOUR BODY'S INNER CYCLES AND RHYTHMS AND LEARN THE BEST TIMES FOR:

- CREATIVITY
- EXERCISE
- SEX
- SLEEP
- AND MORE

CAROL ORLOCK

A Citadel Press Book
Published by Carol Publishing Group

Carol Publishing Group Edition, 1995

Previously published as *Inner Time*

A Citadel Press Book
Published by Carol Publishing Group
Citadel Press is a registered trademark of Carol Communications, Inc.

Editorial Offices: 600 Madison Avenue, New York, NY 10022
Sales & Distribution Offices: 120 Enterprise Avenue, Secaucus, NJ 07094
In Canada: Canadian Manda Group, One Atlantic Avenue, Suite 105
Toronto, Ontario, M6K 3E7

Queries regarding rights and permissions should be addressed to:
Carol Publishing Group, 600 Madison Avenue, New York, NY 10022

Manufactured in the United States of America
10 9 8 7 6 5 4 3 2 1

Carol Publishing Group books are available at special discounts
for bulk purchases, sales promotions, fund raising, or
educational purposes. Special editions can also be created to
specifications. For details contact: Special Sales Department,
Carol Publishing Group, 120 Enterprise Ave., Secaucus, NJ 07094

Library of Congress Cataloging-in-Publication Data

Orlock, Carol.
    Know your body clock / Carol Orlock.
        p. cm.
    "A Citadel Press book."
    ISBN 0-8065-1703-4 (pbk.)
    1. Chronobiology. I. Title.
QP84.6.0745   1995
612'.022—dc20   95-19772
                                                      CIP

For Diane, Maxine, Nancy and Peter, and for Leah and Susan

Contents

I

Introduction

1

Newly Watched Clocks

T his is a tale about time, about the pacemakers we carry inside of our bodies. Our inner clocks keep a different sort of time from the seconds, minutes and hours our watches report. They measure out our lives in increments from nanoseconds to ninety-minute blocks, from weeks, months and years to the longest human cycle, a lifetime. Although we seldom bother consulting these clocks, they began keeping time inside us even before we were born, and they will stop only when our hearts tick their last beats. Since eons ago, back when our species evolved enough to qualify as a species, these pacemakers have been part of us. Yet only in recent decades have we begun learning how to read their faces.

To begin looking at our inner clocks, it's worth doing a bit of time travel ourselves, going back to the dawn of human life.

Picture a clearing near the opening to a cave, forty thousand years ago. A small band of creatures—similar to animals, but nearly human—gathers at dusk. They have been up since dawn, awakened by the sunlight outlining the entrance of their cave. They have lived in this place long enough to notice that sometimes the days seem longer, and at other times shorter. They wonder at this, but not much. Today was a long day, and it was warm. We would call this a summer evening.

Some of these creatures spent the early part of the day hunting their own evolutionary relatives—the bison and hyena and antelope. Others worked quietly, cleaning skins or gathering plants. Now they are active again, talking, preparing beds, collecting wood, then cooking food. While they eat, the sun begins to throw shadows. Before long, one will yawn, and then another and another. This ripple of sleepiness drives them toward the mouth of the cave.

They settle to sleep, their breathing and heartbeats, the separate sounds of their digestion and restless movements combining like the ticking of an immense slow clock. Perhaps one stirs and awakens quite late.

He, or perhaps she, struggles upright—a fairly new achievement for the species. He steps outside to cast a moonlit shadow over the banked fire. This creature wonders why he so often awakens in the darkest hours. It is something only he does, but he remembers that his mother used to do it too. She died in the cooling season, the time when most deaths come, unless an animal brings them. She died during the afternoon, which is unusual too, since most deaths come in the night.

The creature shakes his head and pauses to watch an ember fade. He thinks about time, but in a different way than we think of it today. He knows nothing of minutes, hours nor any division smaller than the unit of a day. Beyond that lie the moon's comings and goings and perhaps the circling of the year.

For him time is a wheel shaped by things that repeat themselves— the migrations of animals, the appearance of certain plants, the tastes of familiar foods. Perhaps he wonders if many small cycles could ever join, as small animals' skins come together into large blankets, making a span that unfurls toward a future.

He kicks a bit of dust toward the fire and watches the ember wink out. There are some things, he supposes, that will never be understood. He pads back into the cave and settles beneath the still-warm skin that makes his bed. His breathing slows and he is asleep.

While our ancestor sleeps, we can fast-forward across time, transporting ourselves to the end of the twentieth century. We might arrive in the contemporary United States, at an address on a quiet suburban street. There, another family sleeps.

Its four members enjoy a comfortable life, a pleasant home and good food. In fact, they have much that their ancient ancestor lacked, except for one thing—time. Like most Americans, these people live at a breakneck pace.

What with two jobs—both the husband and wife work—and two

children—a boy of twelve and a girl of fifteen—plus two mortgages—they're buying a vacation home—it's all this family can do to stay sane and on schedule. Matters aren't helped by the fact that, even while they are sleeping, this family represents a complex jumble of times. Because they are descendants of that ancient, night-waking ancestor, each has a unique biological rhythm. A closer look shows what this genetic inheritance means.

When the sun rises around 6:00 A.M., the father pops awake. As he rolls out of bed, his wife rolls over and burrows deeper into her pillow. She will sleep for another hour, although she'd rather snooze until ten. Her husband pads downstairs and puts on two pots of coffee—decaf for himself, killer-strength caffeine for his wife.

It used to be that their daughter, now a ninth grader, would also be up at this hour. In the past year or so, she's picked up her mother's late-sleeping habits. On weekends the fifteen-year-old is dead to the world until noon. The son has always taken after his mother, so his father sets to work waking him first—a task fraught with complaint—then he knocks on his daughter's door. Soon the coffee's perked, the shower's steaming and doors are banging. All this wakes Mom.

The father gets the kids dressed and dispatched to school. He hates this duty, but his wife seldom remembers to give them lunch money. He arrives at the office early to plow through the morning's work with remarkable efficiency, attend two meetings, and join a client for lunch. After lunch, however, his energy nose-dives. He ate lightly and did not have a drink, but exhaustion always seems to hit at this hour. He blames the pasta salad and proceeds to slog through the afternoon.

His wife, meanwhile, is perking up. All morning her mind felt like soggy cardboard, but after lunch she begins hitting her stride. By four o'clock she's thinking that tonight would be a good time to take the kids out to dinner, then to a movie. She arrives home to find that her husband has returned from work early; he wasn't getting much done anyway. Now he's turned into a couch potato. The idea of dinner out and a movie evokes a sigh that expands to a yawn. She puts hamburger patties in the microwave to defrost.

By the time dinner arrives, both children are home. The twelve-year-old is annoyed because he played baseball at recess this morning and missed two fly balls. He likes Little League better, when games are played in the evening. Three times at Little League he's hit home runs. His sister is also irritated. She wants to visit a friend tonight, but it's a school night. Her parents point out that lately she's been all wound up at bedtime, hardly getting to sleep before midnight, and going out will only make it worse. After dinner she heads for her room to pout and play CDs.

By nine-thirty the father has tuned the TV to a police show and fallen asleep even before anyone has gotten shot. Meanwhile, his wife did two loads of laundry, changed the sheets on the kids' beds and got the youngest into his. She turns on the home computer and begins work on a report she brought home. Her husband's snoring finally overpowers the television's audio, so she rouses him and sends him up to bed. She settles back to work, accompanied by a silent house except for the bass beat of her daughter's music.

As she puts the final touches on her work around midnight, she gets to thinking about time. The whole family is too busy, she realizes, and imagines how pleasant it would be to take a weekend off for a camping trip. They could hike all day, then cook over a fire, sitting and talking while the embers faded. They could stay up watching the stars, then sleep together under them. If only there were time.

How quickly it goes, she realizes. She grew up seeing time as the patient sweep of a second hand around a schoolroom clock. The clock was round, but time was an arrow pointing toward recess, then summer vacation, then graduation. Lately time has turned into a streaking neon arrow.

Of course, these days everything's digital. The kids probably think of time as a series of blips, a sort of ongoing light show.

Her reverie is broken by the bass beat coming from her daughter's room. She heads upstairs to pull the CD player's plug. She herself retires around one, but tosses and turns for an hour, wondering why her husband is always too tired whenever she wants to talk with him. Could he be sick? Is he avoiding her?

At last, like generations of her ancestors, this modern woman sleeps. At her bedside, lighted numbers on a clock dial move through the hours.

A New History of Time

This imaginary household, with its mismatched schedules and missed chances for communication, illustrates what happens in many contemporary families. In part this stems from the accelerated pace of today's world. Yet for these people, as for all of us, age-old physical realities also dictate daily rhythms. Like that night-waking ancestor, we are all programmed to live on individual schedules.

The mother could try to wake up earlier and have time to talk with her husband. Or he might stay up later. Chances are that if they tried this, neither would be on the same wavelength as the other. No matter how hard they work at it, these two will never be at their best at the same time of day.

Similarly, their daughter might turn off her music at a reasonable hour, head for bed and hope to get up earlier. She would probably lie awake for hours, then have her usual trouble the next morning. As for the Little Leaguer, in the afternoon and evening his reaction time and hand-eye coordination are at their best for the day. No matter how hard he practices, he will never hit as many home runs in the morning as he can later on.

This family, like many households, is a patchwork of rhythms. Our rhythms are flexible enough that we can live comfortably with them, and occasionally squeeze in activities around them, but they cannot be changed. Just as some people are taller and some have brown eyes or blue, our individual genetic makeup determines our biological rhythms.

The presence of such rhythms has provoked a chorus of questions over the centuries. Why is it, observers wondered, that some humans sleep more or less than others? Why do some people always awaken late, while others are so lively in the morning? What makes us more alert and quicker to learn at certain hours of the day? Why are we more likely to make mistakes at other hours? Why are we sometimes in the mood for love and at other times unwilling to bother? What inner changes make us fall ill or recover?

Only in the 1950s did answers to these questions begin to emerge. By then technology had led to the design of dozens of precise medical tools—heart and lung monitors, delicate assays of chemical change, intricate and exacting machines for the examination of every feature of metabolism, growth, injury and illness. In 1952 researchers at the University of Minnesota tried out a few of these tools on ancient questions about cycles in living beings. Before long they knew that they had hit a scientific jackpot. They had happened upon an entirely new field of science—chronobiology.

Chronobiology reveals how life—the biology of all living things—

inscribes patterns in time. Every human being, chronobiologists are discovering, is a finely tuned system of many interlocking inner clocks, each displaying a distinct rhythm. Because the concept of time is built into us at the cellular level, biological clocks in our bodies and brains dictate our desires and disinclinations, moods, hungers, abilities and vulnerabilities.

In the decades since that new science began, researchers around the world have begun answering questions about why our ancient ancestor woke each night, as well as why his descendant craved that pasta salad at his twentieth-century lunch. Applications from this new science are arriving in nearly every realm of modern life, from architecture and engineering to law and the study of history.

Physical cycles affect how well politicians and business people compete in our twenty-four-hour-a-day global environment. They also determine how well astronauts survive in the sunless world of outer space. Chronobiological adjustments alter the course of mental illnesses, and they also help athletes peak on the day of the game, enable students to learn more efficiently, and assist childless couples in starting families.

By working with these rhythms, we can protect our health, improve our professional productivity and better understand both ourselves and others. When our cycles go awry, however, both our bodies and our minds suffer. Ignoring them makes us less efficient, and it also puts us at risk for disease and mental illness.

In recent decades this new science has begun bringing ancient mysteries into sunlight and understanding. It is also suggesting the depths of darkness surrounding what we do not yet know.

2

The Impermanent Present

Future historians looking back on our age will remark that it was certainly a good thing we found out about inner time when we did. We discovered our biological clocks quite literally in the nick of time. This is so because we're about to hit a shortage of a natural resource that will, by comparison, make the oil embargo look like an energy glut.

Of all resources, time is the most precious. If a corporation could patent the manufacture of time, the ink on the first advertisements would hardly dry before the government stepped in to regulate this essential product. Like heat, light and water, time is a utility everyone needs.

Everyone also wants it in greater supply, but we cannot mine it, grow it, synthesize it or recycle it. Unlike oil, there are no new wells to be discovered. Time is our least renewable resource.

And it is the resource disappearing most quickly as we accelerate into the twenty-first century. Time's scarcity is already evident in the number of businesses open around the clock, of workers employed on nonstandard shifts or straddling two jobs, and of traffic accidents linked to sleepy drivers—as many as two hundred thousand annually. Symptoms of our time shortage appear in the booming market for time-saving devices—beepers and cellular phones, fax machines, cash machines, overnight mail, and computer networks that can deliver a

9

world's worth of data to our homes twenty-four hours a day. As our supply of time shrinks, we scramble for ways to conserve it; sleep deprivation becomes one of our most pervasive health problems.

Meanwhile, science has discovered that time is limited not only in society, but inside the human body as well. To see how this occurs, we can look once again at that mother, father and two children living in the late twentieth century.

Larks and Owls

In the pleasantly unscientific terminology of chronobiology, this modern family is made up of "larks" and "owls." The two biological types represent separate formats for what different bodies do at different hours.

The father is an extreme lark. In the early morning hours his body's temperature rises, so he wakes up at dawn. His thinking abilities are at their best several hours before his wife's are, but by midafternoon his temperature hits its peak and his efficiency declines. As the sun sets, chemicals begin flowing into his blood telling him it's time for sleep. If he tries to work in the evening, his body will secrete more stress hormones than his wife's will. This man is what many people call a morning person, with his body's clocks running hours ahead of his wife's.

His wife is an extreme owl, or night person. It's a good thing these two got acquainted over lunches, or they might never have found an hour to appreciate one another. Her temperature rises much more slowly than her husband's does, so she sleeps late. That temperature reaches its peak sometime in the evening, accounting for her ability to work far into the night. Her body actually would prefer to live on twenty-six-hour days; she's seldom ready to go to sleep when it's bedtime. For the present, she and her daughter have that much in common.

The daughter, however, is moving beyond her mother's twenty-six-hour day. In a year or two this high schooler may actually function best on twenty-seven- or twenty-eight-hour days. She used to be a lark like her father, but now her body is slipping out of sync with the twenty-four-hour world. Like most teens, she needs as much sleep as ever, but her biological rhythms run on longer days. On weekends when she sleeps until noon, she probably feels her best.

Even their son, still on the twenty-four-hour cycle natural to his age, is feeling the impact of biological rhythms. He inherited his mother's genetic makeup, and he'll be an owl all his life. If his baseball game had been scheduled later, he might have caught those two fly balls. As for the home runs, most athletes perform better late in their biological days. Owls can hit their peak as much as five hours later than larks, so this boy might become a real star if only the Little League scheduled games at midnight.

Larks and owls are two genetically different human beings. We are born destined to become one or the other, and our children inherit either parent's biological cycle or some combination of the two. Most people's daily rhythms overlap fairly well, but twenty percent of us are extreme larks and another twenty percent extreme owls.

As we age, our biological clocks tend to shift us toward larkness. Thus, older people often go to bed earlier but rise with the sun. By the time they're retired and ready to use that vacation home more, this husband and wife will probably have similar schedules.

Wheels Within Wheels

The temperature cycle, determining when we feel most wide awake, and the sleep-wake cycle, telling us when to nod off, are only two of a complex linkage of many clocks that mark our inner time. Other clocks determine when we're hungry and thirsty, when our senses of hearing and smell are sharpest, when our allergies or arthritis act up, and even when we can best lose weight. Some rhythms, like the temperature cycle, affect our whole bodies. Others play out their rhythms in smaller areas—from the brief patterns in cell division to much larger changes in our rate of respiration. Chronobiologists have already located more than one hundred internal cycles, and they're becoming quite casual about new finds. These days folks would sit up and take notice only if they found a physical function that lacked cycles.

To study these rhythms, scientists often go to extremes. They have sealed volunteers, and occasionally themselves, in underground caves, World War II bunkers and laboratories where the light could be kept at a constant level for days and weeks. They have looked at the seasonality of various diseases at different latitudes and examined twins who were reared apart. They have flown around the world on airplanes while checking their blood pressure at hourly intervals, attempted to adapt to forty-eight-hour days, and even attached so-called "vampires" to research subjects, sampling their blood constantly.

Animals, too, have done their part to demonstrate chronobiological patterns. Rats, lizards, cows, crabs and monkeys aplenty have yielded up the secrets of their inner rhythms. Algae display cycles and so do butterflies. Birds definitely keep time, even chickens. In fact, chickens participated in one of the more charming experiments ever done to examine biological rhythms.

This occurred when scientists at a university in India gave some three-day-old chickens a serious case of jet lag. By adjusting the lighting schedule, the Indian researchers arranged for some of their flock to experience repeated eastward travel—losing a few hours a day—and made others undergo the equivalent of westward jet flight—gaining hours. The birds on the eastbound schedule became healthier and grew faster than members of the westbound flock. They even did better than the birds who experienced no jet lag at all.

Fortunately, such results can be obtained merely by turning the lights on and off at prescribed intervals. Otherwise, ambitious poultry farmers might start lobbying for a frequent-fryer plan.

Higemous, Hogemous, Is It Endogeneous?

In any such experiment, the answer to one crucial question tells whether or not a true biological cycle has been found. Is the subject under study—whether chicken, moth or man—actually exhibiting rhythms that originate inside its body? Or are the observed cycles resulting from some rhythmic signal arriving from outside?

It's easy to mistake the response to an outside signal for evidence of an inner cycle. We know, for example, that our hearts beat in regular patterns whether it's day or night, warm or cold, humid or dry. It's far less easy to determine whether our temperature declines overnight because we have a temperature cycle or because the temperature of the surrounding air drops. Like Pavlovian subjects, do we get hungry because the dinner bell rings? Or would we be hungry at certain hours of our biological days even if we could hear no bells and see no clocks? In short, would that biological clock still tick if we were "free-running" on our own?

To find such free-running clocks, isolation is crucial. Chronobiologists must screen out patterns of natural light and darkness, temperature, sound, food availability, social activity and much more. A signal as subtle as a rhythmic change in barometric pressure can violate the pure isolation needed to show whether a rhythm is coming from inside or outside.

One recent experiment illustrates this challenge. In 1989 Stefania Follini, an Italian volunteer, spent 130 days alone in a cave thirty feet below ground in the New Mexico desert. She lived in a nine-by-twelve-foot space, isolated from both sunlight and darkness. The lights in the cave could be turned lower, but never completely off, so Follini used her own body's rhythms to gauge the lengths of her days.

Entering the cave in January, Follini took along four hundred books, a computer, clothing and an array of equipment to monitor her body and its changes. She wore a blood pressure cuff and took her temperature four times within each period she supposed was a day. Occasionally she also used an electroencephalograph to record her brain waves. Each time she guessed a few weeks had passed, she drew a blood sample. She sent that sample, along with other samples and records she had collected, through a tube to colleagues above ground.

During her lonely sojourn Follini's only friends were two mice, whom she named Giuseppe and Nineta. She fed them, and also stayed busy reading books, writing in her diary and monitoring her body.

Gradually she lost track of time. At first she shifted to a cycle on which she slept once in about every twenty-eight hours. Then, after six weeks below ground, her body adopted a forty-four-hour day, still sleeping about a third of the time, or fourteen hours at a stretch. Her perception of how long events took also changed. On one occasion a journalist interviewed her, by computer linkup, for seven and a half hours. She thought the on-line conversation lasted only an hour.

On May 23 Follini emerged to greet her research team and a small band of reporters. She had been underground for more than four months. She thought only two months had gone by.

While she was in her cave, all outside signals that might have cued Follini's inner clocks had been eliminated. These *exogenous* signals masked true inner cycles and would have confused research results. Any cycles that persisted during her time below ground must have come from inside the body. They were *endogenous*, literally ''originating from within.''

Follini's stay in isolation, like isolation studies by colleagues before her, provided a wealth of information about her endogenous biological cycles. Those 130 days delivered data that would fuel the work of medical researchers, psychologists, aerospace scientists and even philosophers for years to come.

Unreliable Clocks

One feature of biological cycles which the New Mexico experiment confirmed was that inner time is far less accurate than the kind of time clocks keep. Irregular cycles, such as the way in which Follini's sleep-wake cycle first shifted to twenty-eight hours and then to forty-four, present a special challenge for those who chart biological rhythms.

If rhythms were perfectly regular, finding them would be fairly straightforward. Some inner cycles are fairly regular, but many are not. Within a given day they may drift a few seconds, minutes or even hours. They will differ slightly depending on the genetic inheritance of an individual and on his or her age. Thus, the periods of various functions—how long it takes them to complete one cycle and get back to the same point—must be based on averages.

Our imaginary family can illustrate this, but only if we ask a lot of them. They must consent to living as Follini did, enduring constant lighting, wearing blood pressure cuffs and perhaps blowing up balloons every so often while being timed by stopwatches. Chronobiological research is no picnic, but when we look at the data, we would notice consistent trends.

We might see that the mother's natural twenty-six-hour cycle of waking and sleeping is irregular. On some days she feels sleepy twenty minutes earlier; on others she heads for bed ten or twelve minutes later. On average, however, she goes to sleep every twenty-six hours. The father's lung capacity might vary on twenty-four-hour cycles, give or take 3.5 minutes; and their son's batting practice—performed hourly—might put him in the cellar from 8:00 A.M. to 5:00 P.M., then qualify him for the all-star game anywhere between 5:30 and 6:15. His sister would get hungry and head for the underground refrigerator every eighty-five to ninety-six minutes. If she often gave in to evening cravings—based on her unique sleep-wake cycle—she would gain weight. If she ate heartily only shortly after waking, however, she would shed pounds.

From rough averages, reliable patterns would emerge. Yet because their internal clocks are free-running down in that cave, the members of this family would also drift away from one another even more than they do now. While participating in a similar experiment, three men got so out of stride with one another that when they once happened to sit down to eat at the same time—a rare occurrence, since all three were operating on different subjective "days"—one ate breakfast, a second ate lunch, and the third ate dinner.

In addition, the data from our experiment would reveal that not all biological cycles move through their periods gradually. The father's temperature might climb and decline slowly each day; but the levels of hormones in his blood might leap up and down, since hormones are released into the blood in pulses—a sort of all-or-nothing arrangement. If we drew a chart showing this man's temperature pattern, our picture would resemble smooth, rolling waves. A chart of his hormones would square off those waves into a blocklike pattern. Charts of other cycles might show regular zigzags or irregular spikes, pictures of crashing waves or jagged mountain peaks.

Thus, within each individual, many cycles go on constantly. In a sense, the inside of our bodies resembles the showroom of a master clockmaker, a vast interior where many timepieces contribute to a great ticking uproar. These play out a complex symphony, some cycles overlapping, some driving others, some operating independently.

The Real Clockmaker

No human clockmaker built our biological clocks, of course. Instead, many forces combined to shape them and link them together. Our inner cycles represent life's adaptation to a world that is, itself, constantly cycling.

We evolved from animals whose environment kept changing, but doing so in regular cycles. The earth rotates on its axis every 23 hours and 56 minutes. Every 29½ days, the moon revolves around the earth. Meanwhile, the earth takes 365 days and 6 hours to revolve around the sun. In nature, nothing stands still.

Amid such cycles, a given locale gets a daily average of 12 hours of light and 12 of darkness. That's a very rough average, though, since the actual amount depends on the latitude and the local season. The seasons also determine whether this changing environment is warm or cold, wet or dry, lighter or darker. In addition, on some nights moonlight makes the darkness less dark.

To survive, animals needed to awaken and feel hungry when food was around, sleep long enough to stay hidden and safe from predators, and be ready to mate when the conditions would be right for their offspring. In chronobiological terms, "Adapt or die" meant "Cycle or die." Our bodies still carry the remnants of nature's harsher realities.

Clues pointing to our inner cycles were evident from the time science began studying plants and animals, but it took a long time for anyone to recognize the pattern. Evidence has been popping up since before the dawn of history. Unfortunately, due to a series of historical accidents and oversights, no Sherlock Holmes arrived to piece the puzzle together.

To understand why it took so long, as well as to appreciate the new discoveries arriving every day, we need to begin at the beginning. What we know about chronobiology came not only from isolation experiments and jet-lagged chickens, but in the course of a history so quirky that, were it not true, it could only have been invented by a collaboration between Woody Allen and Salvador Dali.

II

What We Wondered

3

In a Time Before Clocks

C hronobiology today is most like the classic discovery scene in a
mystery story: The detective calls all the suspects together in the
drawing room, ticks off the clues on his fingers, and then fingers the
suspect. But the clues that would reveal chronobiology's mystery go
much further back than most who-done-it plots.

For at least thirty thousand years our species has been interested in
how time works. Ancient cultures celebrated the new moon and held
ceremonies at summer and winter solstices. People supposed the moon
had an impact on women's menstrual cycles and noted how increases
in sunlight affected plants' productivity, animals' fertility and—
presumably—their own fertility as well. They praised deities with
names like Helios, Luna and Hyperion—the sun, the moon and dawn.

Just as time ruled religion, it also became a part of medicine. In the
East, Chinese and Tibetan medicine prescribed drugs based not only
on the time of day but also on the day of the week, the monthly
menstrual cycle, and even the season of the year. In the West, the
Greek philosopher Socrates noted that the symptoms of mental illness
worsen during certain times of year, and Hippocrates, a physician,
confirmed that the solstices and equinoxes are the most dangerous
periods for emotional symptoms. His were among the earliest observations of a syndrome that was medically recognized only in the 1980s.
We now call it Seasonal Affective Disorder, or SAD.

Hippocrates also observed rhythms in physical illnesses. He noted that respiratory infections occur far more frequently in winter, and that changes in his asthmatic patients' symptoms followed twenty-four-hour cycles. When it came to treatment, Greek physicians used their knowledge of biological rhythms to prescribe medicines in carefully timed cycles, usually of seven days each.

While Greek medicine refined its awareness of time, philosophers and naturalists added observations about rhythmic cycles in other creatures. Aristotle remarked on the fact that the ovaries of the sea urchin, a relative of the starfish, swell up during the full moon, then become smaller the rest of the month. Another observer noted that the leaves of the tamarind tree open in daylight, then close again at night.

People who read this report may have supposed that the tamarind's leaves reflected feelings similar to those that Greek myth attributed to heliotropes. These plants' daily changes of leaf position gave them their name, which means "turning toward the sun." Supposedly, heliotropes' leaves expressed these plants' unrequited love for the Greek sun god as Helios rode his chariot across the sky.

Meanwhile, more clues surfaced. An anatomist used a water clock to determine that our pulse rates rise and fall depending on the time of the day. The Roman philosopher Cicero remarked that the availability of oysters and other shellfish depended on the phase of the moon.

Thus, well before the birth of Christ, those who studied plants, animals and ourselves were curious about rhythmic cycles in living systems. Yet it would take more than two thousand years before researchers looked closely at these phenomena. For a variety of reasons, reports about seasonal madness, asthma, plants' movements and sea creatures' monthly cycles fell by the wayside in history.

Why Inner Time Waited

With twenty-twenty hindsight, we might wonder why no one took on the job of studying the many sightings of rhythms in biological systems. History zigzags, however, and the path to our understanding of chronobiology had more zigs and zags than a spool of rickrack after a cat has bested it in a wrestling match.

For starters, researchers faced a shortage of accurate tools. A curious scientist could not closely check the schedule of his plants' movements or his patients' symptoms, because he lacked any clocks. Until the 1600s he had to glance out the window at one of the few available mechanical clocks, usually on a church tower or in the town square. He could hardly expect to follow the sweep of a second hand—there were none—and candlelight, which burned brightly until Edison blew it out in 1879, made it difficult to monitor a patient's symptoms twenty-four hours a day.

In addition, while clues to the mystery of our body's rhythms kept cropping up, those who knew about them remained hidden from one another. Related findings came from widely diverse fields, but who would think to connect the opening and closing patterns of a flower's petals with the classic twenty-four-hour pattern of a fever or with an earthworm's movements? No one did, and nature kept its secrets.

Considering the disadvantages they faced, it's surprising our ancestors learned as much as they did about rhythms. Yet in the years between 1300—when the Chinese looked at time and estimated that evolution had required 130,000 years—and the mid-1800s—when Charles Darwin charted his own version of evolution—botanists, naturalists and physicians kept turning up clues about biological cycles. Even one explorer contributed.

Finding False Horizons

On the night of October 11, 1492, Christopher Columbus steered the *Santa Maria* through a moonless night toward what he conjectured would be the Far East. He missed his destination by roughly seven thousand miles, but happened upon a useful land mass in between, as well as a curious fact about biological rhythms. Yet here, too, Columbus succumbed to optimistic misjudgment.

Around 10:00 P.M. that night, he noticed light out where the horizon ought to be. This must be a sighting of land, he assumed since land birds had been overflying the *Santa Maria* all day. The light waxed and waned, flickering like a guttering candle. Columbus could not be sure, and it was only later that another sailor, aboard the faster *Pinta*, actually sighted land. Nonetheless, Columbus was onto something.

As it turned out, he must have seen the mating of Atlantic fireworms, marine worms with a highly structured sex life and a fondness for the waters around the Bahamas. Fireworms discharge their luminous streams of eggs and secretions onto the water only once a month, precisely one hour before moonrise on the night before the moon enters its fourth quarter. October 11, 1492, was just such a night, and as Columbus stood squinting for a sure sighting of land, the moon rose precisely one hour later.

It took five centuries for a marine biologist to identify the cause of Columbus's luminous vision. A similar fate awaited many insights about rhythms in nature. In the 1600s the French philosopher René Descartes called the pineal gland the "seat of the soul," and that description was apt enough to suffice for three hundred years. Only in the twentieth century did scientists recognize that the pineal influences our bodies' cycles of alertness and mood. The English writer Robert Burton wrote, in 1628, "Our body is like a clock; if one wheel be amiss, all the rest are disordered." He could not have expected to wait two hundred years before anyone caught on to what he meant. Only in 1814 did the Frenchman J. J. Virey describe biological rhythms with the phrase *horloges vivante*—literally, "living clocks." Even Virey's understanding would gather dust for another century and a half before being taken seriously.

In the meantime, a handful of researchers pressed beyond flickering insights to shed serious light on these *horloges vivante*. The first significant study came from Paris in 1729.

Clocks That Grow

The astronomer Jean Jacques d'Ortous de Mairan happened to own a heliotrope, that plant whose love of the sun Greek myth had celebrated. Its daily leaf movements—toward the sun all day, folded closed at night—intrigued de Mairan, and he got to wondering what might happen if his plant went a few days without seeing its beloved sun. Would lonely leaves stay folded up? The unsentimental scientist locked his plant in a dark closet. He peeked in every so often and discovered that apparently the Greek sun god also made visits to the closet.

Like clockwork, the heliotrope dutifully opened its leaves when it was daylight outside and closed them when night arrived. De Mairan hazarded a guess that plants had internal timekeepers, capable of functioning in the absence of sunlight.

As a scientist, he immediately foresaw the possibilities. To demonstrate the existence of plants' inner clocks, one might put them in darkness all day, then under candlelight at night. Tests could be made on other plants, and on whether temperature changes were the true cause of the leaf movements. If the internal timekeeper did exist, it might explain the uncanny ability of bedridden invalids, who were often kept in darkened rooms, to awaken when it was daylight outside and to sleep at night.

Unfortunately, de Mairan was a busy man. Preoccupied with lofty matters of astronomy, he hardly thought this observation deserved mention. Because a friend insisted, he did allow a report to be submitted to the *Proceedings* of the Royal Academy of Paris, but apparently the *Proceedings'* readers were busy too. Thirty years passed before anyone bothered following up.

That researcher also lived in Paris, and when Henri-Louis Duhamel du Monceau saw de Mairan's report, he probably gave it his best Parisian sneer. De Mairan must have been careless, du Monceau assumed. Light surely got to that heliotrope. To disprove the report, he carried his own plant into a cave one August morning. He came back the next day at 10 A.M. The leaves, which he expected to find folded after twenty-four hours in total darkness, stood as open and erect as if the early morning sun were shining on them.

Du Monceau gave his plant a few more days to disprove de Mairan. The plant continued to defy his expectations, so late one afternoon he carried it, with its leaves neatly erect, out into sunlight again. That night a strange thing happened. The leaves, which should have closed when darkness fell, stayed wide open for more than twenty-four hours. The following night they resumed their regular rhythm and kept with it.

We now know that the pulse of sunlight which du Monceau's plant received late in the afternoon caused it to become confused about its timing. Such shifts in the expected timing of light can alter human patterns too. Du Monceau had no such understanding, of course. Yet he remained unconvinced about de Mairan's report. There simply must have been light in the cave.

Du Monceau took a plant and locked it in a trunk. He smothered the trunk in heavy wool blankets. He trundled the whole assembly in a closet. Du Monceau was very determined, but so was the plant, and when he opened the closet, tossed off the blankets and snapped the locks, he found leaves keeping perfect time.

In such cases our natural tendency is to root for the plant, and it did emerge victorious, although not before further battle. First du Mon-

ceau had to disprove his suspicion that daily changes in temperature
were signaling the leaves to open and close. He headed out to his
hothouse, cranked up the stoves full blast and watched leaf movements
overnight. The plant's clock hung in there, keeping right on ticking,
oblivious to the abuse. Bested, du Monceau now allowed that "the
movements of the sensitive plant are dependent neither on the light nor
on the heat." It was as much as he would grant.

Meanwhile, farther north, a Swedish botanist was taking a different
approach to plants' daily movements. Carolus Linnaeus, at work
classifying plants and animals into the system still in use today,
devised a novel way of illustrating plants' uncanny accuracy as
timekeepers.

Instead of constructing experiments, Linnaeus created a garden
outside his window and used it as a timepiece. Depending on whether
the lily's petals were open, or whether the scarlet pimpernel lay closed,
Linnaeus could judge the hour any time between 6:00 A.M. and 6 P.M.
in the evening. Linnaeus's flower-clock was a nifty system, ecologi-
cally correct, and very smart in a day when most clocks were still
unreliable.

Clocks did eventually gain greater accuracy. Pocket watches became popular, thanks to the helpful innovation of minute hands.
Thermometers also improved. Before long, one doctor noted that the
body's temperature rises and falls 1.5 degrees over the course of the
day. Another observed that all diseases display twenty-four-hour cycles. A third physician recommended that opium, often used as a
painkiller, be given in the evening rather than the morning to achieve
the best effect. With so much attention being paid to time—in an age,
in fact, when nearby towns designated themselves as having separate
time zones—it was not surprising that some of the finest scientific
minds began investigating questions about biological timing.

The Accuracy of Inner Time

Those two Frenchmen, de Mairan and du Monceau, had confirmed
that plants had biological clocks, but neither bothered checking on
whether those clocks were very accurate. It took a Swiss scientist,
possessing the advantage of better clocks, to do that. In the process, he
added a new angle to our perception of inner timekeepers.

The botanist Augustin de Candolle confirmed that plants' leaves did indeed open and close according to the time of day, and he also determined that changes in humidity did not affect these movements. But what would happen, he wondered, if the plants were exposed to longer light, say for a full twenty-four hours? De Candolle set up six lamps and kept them burning over some mimosas all night.

When darkness fell outside, the plants dutifully went to sleep. When day arrived outside, their leaves opened again. Under continuous light, however, the mimosas appeared to reckon that night came every twenty-two or twenty-three hours, rather than every twenty-four. It seemed that under lights, plants' clocks ran a mite fast.

Next de Candolle tried reversing day and night, keeping his lamps going all night but leaving plants in the dark during sunlit hours. Soon leaves were turning every which way, without reference to the clock hour. After a few days they stabilized, but in a rhythm that treated their lamp-lit night as day and their darkened day as true night. De Candolle had succeeded in provoking plant insomnia, reversing his mimosas' sleep cycles by 180 degrees.

De Candolle went on to check the clocks of other plants and found that almost all had inner clocks. He hazarded a guess that different species would need different levels of brightness to control their timekeepers, but, like so many before him, he never followed up on his hunch.

Only in our own century would science confirm that not only plants but animals and humans can free-run when exposed to continuous light or darkness. Under such conditions, living creatures shorten or lengthen their usual twenty-four-hour cycles. De Candolle's hunch about the timing differences between different species also turned out to be true. Understanding why, however, would require complex knowledge of both evolution and genetics.

The Logic of Living Clockwork

The bombshell of evolutionary theory, *Origin of the Species*, was published in 1859. It would make Charles Darwin's name a household word, although not always a polite one. Lesser known are Darwin's experiments with plants' daily movements, as well as his curiosity about earthworms. As he approached the age of seventy, ailing and confined to his couch, Darwin had time to watch his window plants. He noticed that they kept moving around. That got him to wondering.

Soon Darwin devised a series of experiments, thousands of them. He exposed plants to light coming from a variety of directions. He boxed them so they could receive light only from above. He used strong light and dim, light that was diffused and light that was focused. He enclosed portions of plants in glass tubes painted with India ink to shield them from light. He used props to force leaves to stay open at night. Curious even about how sound might affect their movements, he arranged for his son to play a bassoon nearby while he studied them.

Darwin's results, published in *The Power of Movement in Plants*, confirmed the work of the early French pioneers and de Candolle, while further refining it. The father of evolutionary theory concluded that plants' movements supported his theory, illustrating adaptation to the availability of light and the dangers of freezing at night.

Before his death two years later, Darwin published only one more work, a small study on the activities of earthworms. These creatures, too, he found, exhibited daily rhythms, spending their days underground and coming out only at night. Had he lived, Darwin might have contributed even more to the modern understanding of biological rhythms, but time itself intervened before timing could be further explored.

4

The Arrival of Indoor Sunlight

T he invention that may have had the greatest impact on our understanding of biological rhythms came from outside the field of
biology. Once Thomas Edison invented the light bulb, our access to
usable time more than doubled. Yet our appreciation of our own
biological rhythms suffered a nearly crippling blow.

To grasp the impact of the light bulb, we must imagine how our
preelectric ancestors lived. Before candles became available, creatures
of our kind, like all diurnal, or "daylight," animals, rose with the sun
and went to sleep once darkness fell. We probably rested longer in
winter and worked harder in summer. Even after candles became
available, good light was hard to come by once the sun set. If one ran
out of paper or grain, it made sense to wait until morning before
heading out to the market or barn.

Oil and gas lamps helped, allowing streets to be lit and household
chores to be delayed into the evening. Yet those who sewed or kept
accounts by such light knew their hours of wakefulness were numbe-
red. Natural light far outshone anything human-made. It seldom
caused eyestrain; it never needed its wick trimmed.

Then in 1879 Thomas Edison introduced a device which, he re-

27

marked, would make electric light so cheap that only the rich would be able to afford candles. Within four years Edison's bulbs were lighting one square block of New York City. Slightly more than a century later, we live in a world where only the rich can afford to limit their work to the hours from sunup to sundown.

The rest of us trust alarm clocks to set our bodies going, often before sunrise. We regularly come home after dark, only to take up household chores. A quarter of us are employed in shift work, with schedules that may require that we sleep while it's light outside. For diurnal animals, it's a peculiar way to live. We may enjoy attending midnight movie marathons, visiting all-night delis, and partying until dawn, but we also battle sleep deprivation, insomnia and the risk of falling asleep on the job.

Were he alive today, Edison would probably scoff at our troubles. He considered sleeping all night a waste of time. He could make do with naps; couldn't everyone? As it turns out, Edison and others like him—Winston Churchill and Albert Einstein among them—got genetically lucky when it came to their biological rhythms. Some people need a lot less sleep than others. As for the rest of us, we may be clock-carrying members of twenty-four-hour society, but we inhabit bodies that were designed to function at full alertness for only about twelve hours at a stretch.

Edison's invention of that steady, all-night light was not the only nineteenth-century setback for our relationship to our biological rhythms. While more clues kept arriving, clues that might have helped us hear our inner clocks ticking, history also heard from theorists who were way off track when it came to reading human clocks.

Mysteries of Timing

The same year Edison turned on the lights, a curious researcher reported that the statistical incidence of suicide was seasonal. Slightly more than a decade later, a French sociologist confirmed the report: Most who took their own lives did so in the springtime—on spring afternoons, to be precise. The statistic didn't make sense, but there it was in the numbers.

Nor did it make a lot of sense to subscribers of *Popular Science* in 1888 to learn that animals could read clocks and calendars. Yet as one letter writer pointed out, after several weeks of giving cows salt on Sunday mornings, he would find them coming in from the fields early—at the appointed hour—only on Sundays, ''standing at the bars, the point nearest the house, with every appearance of mute expectation.'' He eliminated the possibility of nearby road-travelers or other external signals alerting the cows to the day and time, and concluded, ''So far as we could judge, one day was like all days except Sunday, which they might have called *salt day*, had they possessed the faculty of speech.''

As if spring afternoon suicides and clock-watching cows were not enough, nineteenth-century science was coming up with all kinds of strange evidence about human rhythms. Epileptic seizures, one observer noted, were more likely to occur in the morning. Another added the insight that, in women, seizures were likely around the time of menstruation. The likeliest hour for birth, on the other hand, was the dead of night. During those late night hours, as yet another researcher found, the body's temperature declined. It fell to its lowest point in the morning and rose all day to reach an evening peak. Toward the end of the century, a German investigator even reported that the well-known pattern of urine flow in human beings—heavier in the morning and nearly nonexistent at night—could not be altered by altering one's way of life. He stayed in bed and took fluids around the clock. Yet when the sun rose, he was up, too, and headed for the water closet.

Throughout the nineteenth century, theories about biological rhythms flew thick and fast, launched by responsible researchers as well as crackpots and mountebanks. In 1887 the author of one of those theories singlehandedly almost scuttled the study of biological rhythms.

Biorhythm Hokum

Wilhelm Fliess was a German nose and throat specialist. For reasons best lost in the mists of history, Fliess believed that, on the cellular level, every human being is bisexual, and that elements of both the male—strong, courageous—and the female—sensitive, loving—were present in every cell. They expressed themselves in cycles of twenty-three and twenty-eight days, respectively. These elements showed up clearly in the mucus cells lining the nose. Depending on whether these cells' sexual cycles were in sync or out of it, one felt better or worse, both physically and emotionally. As Fliess saw it, nasal irritations were nothing to sneeze at. They indicated neurosis and sexual abnormality.

Today such ideas may strike us as too wild for words, but the nineteenth century took seriously equally oddball ideas about germ theory, radio waves, radioactivity and psychoanalysis. All of those turned out to be useful. As it happens, the father of that last theory, Sigmund Freud, was among those who took Fliess's beliefs to heart.

Perhaps it's best not to inquire what sexual abnormalities troubled Freud, but we do know that the father of psychotherapy had Fliess perform two operations on that distinguished Austrian nose. The procedure required the application of cocaine—known for its anesthetic powers—to the "genital" cells of the nose. Apparently it worked for Freud. It certainly worked for Fliess.

Fliess's medical practice became one of the largest in Berlin. Copies of his books on biological rhythms sold like hotcakes, although one critic described his work as "a masterpiece of Teutonic crackpottery." Fliess's ideas survived long enough to be added to in the 1930s, by a teacher who also postulated a thirty-three-day cycle of creativity or intellect. The adapted theory survives into our day as "biorhythms," the inspiration for imaginative newspaper charts which appear between the astrology column and the comic strips.

The trouble is, as shown through many serious investigations, biorhythms just aren't there. Researchers have repeatedly tried to confirm the theory and repeatedly failed. Biorhythms don't predict baseball players' best batting days. They don't match up with miners' worst days for accidents. They won't figure into workers' productivity, drivers' traffic accidents or airline pilots' likelihood of causing a crash. Researchers soon concluded that biorhythms are sheer bunk.

Among scientists, Fliess's theory was pretty much dead within decades, but because biorhythms became confused in the public mind with true biological rhythms, the damage was done. No serious scientist was about to advance a theory about health or disease based on cycles that resembled biorhythms. The few who tried heard raspberries of ridicule. By driving capable researchers from the field, Fliess's theory—perhaps as much as lack of technology or ignorance of others' efforts—delayed our understanding of the body's cycles.

A Steady State of Change

In the meantime, another theory, one boasting a far better pedigree, led to equal misunderstanding. This was the theory of homeostasis, and it went directly counter to the idea of biorhythms.

Since the time of Hippocrates, physicians had noted that the body's inner workings possessed a quality that the Greeks called harmony. They believed that when our inner systems work in harmony, we enjoy health; and ancient regimes of medication sought to take unharmonious systems and teach them harmony again. As in music, which the Greeks understood well, harmony need not mean every instrument playing at the same tempo. Rhythms could complement one another, both in good music and good health.

And yet, physicians wondered, how could the body maintain its harmony if the heart, after a brisk run, began drumming too fast? How could a system stay in tune if summer heat or winter cold altered its temperature?

Around the end of the nineteenth century, an idea arrived to explain the body's constancy amid changing conditions. Apparently our varying rhythms kept their harmony thanks to an internal orchestra conductor. This conductor enforced a system that later came to be known as homeostasis.

If the heartbeat accelerated, homeostasis brought that drummer back into line. If we got too hot or too cold, homeostasis made us perspire or shiver, gradually leveling out our temperature. In short, the body could adjust when external conditions disrupted its perfect harmony.

There were, of course, variations within the homeostatic state. Medical readings of temperature, blood pressure or hormone levels might differ slightly from one day to the next, or even from one hour to the next. Yet these were merely random blips, destined to be brought back into line by our homeostatic systems. Within specific ranges—recognized as normal—what went on inside us looked pretty constant.

This homeostatic approach has lasted into our own day, providing useful insights into how the body actually works. Yet its explanation of those random blips overlooked the obvious.

We now know that a ''normal'' temperature or blood pressure reading in the morning may in fact be a sign of illness if it shows up at night. Our temperature, blood pressure, hormone levels, heart rate—in fact almost every internal measure—vary significantly during the course of a day.

In addition, by assuming that the body stayed the same all the time, the homeostatic theory assumed that medical treatments would achieve the same effect no matter what time of day they were given. A pill, an exercise regime or a surgery scheduled at any hour would provide the same results. We now know this isn't true. An action as simple as taking an aspirin has a different effect on the body depending on the time of day we do it. By assuming that diverse readings were mere accidents, the concept of homeostasis wrote off clues that might have increased an understanding of our biological rhythms.

5

New Clues to Nature's Clockwork

W hile the Industrial Age matured—with light bulbs burning through the night and fires burning in the bellies of thousands of factories—the age of specialization approached. Insights about biological rhythms began coming from a wide range of experts. Entomologists found rhythms within insects, and agricultural experts found them in plants. Doctors and explorers, mathematicians and psychiatrists, began turning up clues. Yet before the many observations of inner rhythms came together, each scientific detective pondered a maddeningly incomplete piece of the puzzle. Those pieces accumulated slowly, and the puzzle's picture grew—to quote from *Alice in Wonderland*, popular at the time—"curiouser and curiouser."

Insects Got Rhythm

In the early 1900s the physician Auguste Forel was fortunate enough to own a summer home in the Swiss Alps. Out of doors on the verandah, his family enjoyed leisurely breakfasts, including locally made jams and marmalades.

Unfortunately, bees loved the sweet spreads too, and they tended to gather around the table and drive the family indoors. Even when the Forels sat down to breakfast inside, the bees showed up beyond the window.

Forel consulted his watch. He kept an eye out for bee activity on the porch over the rest of the day. Only at the breakfast hour did bees mass on the verandah. They acted as if they expected to be served.

The bees might possess some memory for time, Forel supposed, but he pursued the matter no further. A few years later a German zoologist would echo his guess. Bees, he remarked, visited buckwheat fields only during the hours when plants' blossoms released nectar—between ten and eleven in the morning. He coined the term *Zeitgedächtnis*, suggesting quite literally that bees possessed a "time sense."

It was not long before a young scientist decided to test these ideas about bees' time sense. To do this, Ingeborg Beling, working in 1927, needed to make sure the same bees came on time every day. After enticing a group of them to hold still long enough—she used sugar water—she painted identifying dots on each. Next she trained them to follow a scented trail to a reward of nectar. Soon the marked bees were finding her feeding table reliably, and Beling took a crucial step.

Between 4:00 and 6:00 P.M. she put out nectar. Bees that came to the feeding table at other times went home hungry. Even more quickly than they had learned where to find the nectar, the bees learned what time to look for it. Finally, to make sure that no scent was drawing them, Beling stopped putting out any nectar at all. The feeding station remained fairly quiet all day, but between 3:30 and 6:30 P.M.—since apparently some bees are more punctual than others—there were enough takeoffs and landings to give an air traffic controller a case of hives. Most of those visits came between 4:30 and 5:30, right in the middle of the usual dinner hour.

In further experiments, Beling tried training bees to check for food every nineteen hours or every forty-eight. She met with defeat. Yet even six hundred feet below ground, in the bottom of a salt mine, bees could be trained to seek food at an appointed hour based on a clock that, as it turned out, ran on a 23.4-hour day.

Apparently bees could keep track of time. It was equally apparent that they wore no watches. Was their clock an internal timepiece, or might they be responding to some environmental signal the scientists had yet to identify? The only way to find out was to train bees in one environment—which might be providing subtle time cues—and then transport them to another time zone, providing different cues. If the bees stayed on schedule, they had internal clocks.

Years would pass before this experiment could be tried, but by 1955 airplanes were flying across the Atlantic. On June 13 of that year a researcher boarded a plane carrying the somewhat unusual cargo of forty bees packed in a box.

These were Parisian bees, and they had been trained under controlled conditions in a room that screened out any clues as to external conditions. This bee room also had a twin, an exact replica at the American Museum of Natural History in New York. The twin room awaited the bees' arrival.

After traversing 76 degrees of latitude, the researcher and his bees landed. New York time was five hours different from that in Paris, but when the bee box was opened in the twin room, the trainees headed out to look for food at 3:00 P.M. Eastern Daylight Time, precisely twenty-four hours after their Paris feeding time. Apparently unaware of their new address, the bees had not reset their watches. They could not have used any external signal to keep time, so the bees definitely had internal clocks.

As it turned out, these clocks could run reliably, oblivious to local time, for at least a month if bees stayed in controlled environments. When bees trained in one time zone were freed in another, they needed a couple of days to reset their watches. Later measurements revealed that bees, somewhat like ourselves, can adjust their internal clocks only a couple of hours each day. Nor is it necessary to actually transport them to a new locale in order to reset their clocks. Simply keeping bees in a controlled environment while altering the times the lights go on and off can fool New York bees, or any others, into thinking they live in Paris, Shanghai, or anywhere.

While these scientists were studying bees, others were busy discovering clocks in other insects. By the mid-1950s it was clear that many insects could tell time reliably. Fireflies could accurately schedule their evening light show even if kept in continuous light, and that ancient scourge of modern civilization, the cockroach, had a clock. Careful dissection of the roach's brain threw its daily habits— scurrying around at night, snoozing all day—quite out of whack. In the roach, at least, it began to look as if a clock resided in the optic lobe of the brain. It also appeared that insects' ability to keep time might be inherited.

How Plants Plan Ahead

While entomologists studied insects' clocks, botanists unlocked more secrets about how plants keep time. The inspiration for some of this work came from those thousands of experiments Charles Darwin had done.

In the 1920s a Dutch botanist, Anthonia Kleinhoonte, noticed some questions Darwin had overlooked. Kleinhoonte wondered if plants had clocks even when they were young sprouts. And what would happen if she changed their cycles by using artificial light? She focused on one plant, the jack bean, and started out with seedlings. It took seventeen or eighteen days, Kleinhoonte soon discovered, for jack beans' leaf movements to consistently display twenty-four-hour cycles. Next she waited until the middle of the night, when the plants' leaves were folded in sleep. She flicked the light on and off, then left them in continuous darkness.

As earlier experiments had shown, the leaves eventually opened as if daylight had come, but this daylight came at the wrong time. By flicking on the light, Kleinhoonte had delayed the leaves' opening by twelve hours, precisely as if the plants believed the sun had come up and gone down very quickly. Apparently jack beans were determined to get their allotted sleep before rising, no matter how brief the day had been.

Soon Kleinhoonte was putting her plants through their paces in cycles that hardly resembled normal days. She trained seedlings to behave as if the sun rose and set every sixteen hours, dutifully obedient to lights that went on and off at eight-hour intervals. Trained from the time they were sprouts, such plants had never known a normal twenty-four-hour day. Yet when Kleinhoonte deprived them of their sixteen-hour cycle, leaving them in continual darkness or continual light, the plants abandoned the artificially accelerated day. Instead, they independently invented twenty-four-hour cycles, opening and closing their leaves in a rhythm as regular as the movements of a sun they had never seen.

While Kleinhoonte was raising, and seriously confusing, those jack beans, another researcher developed an interest in a different bean—the scarlet runner. In the early 1930's the German botanist Erwin Bünning posed a question Kleinhoonte had not asked: Would bean plants develop rhythms if they were raised entirely in darkness or in light? Would they know what a twenty-four-hour day was if they'd never seen one?

Bünning let his seedlings grow, some in complete darkness and others in continuing light, until they were mature enough to display rhythms. Yet because the light, or lack of it, stayed the same day and night, the scarlet runners seemed unable to figure out the system. Leaves opened and closed at random. There was no clockwork.

Soon Bünning began to wonder what might happen if the beans saw a change in lighting just once—say by flicking on the light in the utterly dark room, or by briefly switching off the bulb in the light room. The plants answered his question immediately. Having seen the brief change, both sets of plants figured out twenty-four-hour cycles for themselves and proceeded to follow them whether the lights later stayed on or off. Apparently plants caught on fast.

But they could also become forgetful. If plants were left in darkness or light for too many days, their rhythms gradually faded out. They needed another signal to get them ticking again. Nevertheless, as Bünning discovered by crossbreeding different species, plants not only had accurate timekeeping ability, they could also do math. Crossbreeding a plant that had a 24.2-hour cycle with another that cycled every 24.6 hours produced a generation with a cycle of 24.4 hours—precisely the average of the parents' daily rhythms. In 1958 Erwin Bünning published the first major study of the rhythmic movements of plants.

Clockwork Creatures Great and Small

Insects and plants aside, the living creatures for which rhythmic movements had most often been observed were birds. Soon researchers turned to them.

As everyone knew, birds migrated from place to place on schedules as reliable as any calendar. In the early 1920s one Canadian scientist tried changing the migration plans of a flock of migrating juncos.

It was autumn and the birds were headed south, but Vernon Rowan captured some in outdoor aviaries. He kept electric lights shining on them a few minutes longer each day. By mid-December the juncos were singing springtime mating songs. When he finally released them, in midwinter, the exposure to light had turned Rowan's birds into wrong-way juncos. Thinking it was spring, they headed north, probably to a fate too chilling to contemplate.

Within decades, other researchers confirmed that birds chose the date and direction of their migration by the position of the sun. Yet to do so, as everyone recognized, the birds must compensate for the sun's daily movements. Otherwise they'd have a terrible time orienting by a marker that was low in the east in the morning, overhead at noon, and falling in the west at dusk. Granted, migrating birds must have a compass, but could they also possess a clock?

The answer came in the 1950s from work done by a German biologist. Aware that birds took their bearings from the sun's direction, Gustav Kramer caged a starling outdoors and trained it to find food in a feeder on the west side of its cage. He put identical feeders on the north, south and east sides, but only the one on the west offered any goodies.

Although the sun's location changed from morning to evening, the bird could apparently compensate, seeking food by flying away from the sun in the morning and by flying toward it in the evening. Yet how did the bird know the difference between morning and evening? Kramer decided to find out what a bird would do if the sun stood still.

He covered the cage to block out the real sun. Next he provided an unmoving electric sun, installing it on the west side of the cage. In the morning he turned on this artificial sun, making dawn come in the west when it ought to be in the east. Could the bird still figure out which way was west?

Apparently it could not. The dawn was false and came from the wrong direction, but the bird took it for the real thing. It turned its tail feathers toward the unmoving sun and looked for food in the east feeder, going 180 degrees in the wrong direction. As hours passed, the sun did not move but the bird did. By noon it was checking the north feeder, 90 degrees the wrong way. It finally found its customary meal ticket—the west feeder—at six in the evening. Reckoned by the sun's movements, time had stopped; but the bird was counting hours and theorizing about where food ought to be. A bird could tell time.

Like plants before them, birds' daily rhythms tended to go awry after a couple of weeks in complete darkness or under lights burning twenty-four hours a day. They began operating on days that were slightly longer or shorter than usual. Also like plants, birds apparently reset their clocks when the lights went on or off at unexpected times—exposing them to a pulse of light in the dead of night, for example.

They refigured their schedules and went to sleep based on a day that appeared to have begun at the wrong time.

Well-Timed Lab Rats

Birds were fairly handy for experimentation, but they also presented serious drawbacks. Each year they might fly thousands of miles north or south of the latitudes that housed well-equipped laboratories. Animals which stayed closer to home were much easier to keep tabs on, and rats fit the bill perfectly.

Rats may be favorite experimental animals, but perhaps it is just as well that humans have little sympathy for them. As researchers worked to understand their rhythms, rats had their internal clocks spun wildly out of cycle, were starved, dehydrated, blinded, heated, chilled and even subjected to electroconvulsive shocks. Their revenge, should they ever decide to wreak it, would make *Gremlins* look like *Mary Poppins*.

Rats were easy to come by—one experimenter caught his candidates wild around Baltimore. They liked to run on exercise wheels, feed and mate at night, then sleep away the day. Like plants' seedlings, newborn rats required a couple of weeks of growth before they displayed daily rhythms. Also like plants, rats' internal clocks could be thrown out of whack by the manipulation of light and darkness.

By the middle of the century, the American physiologist Curt Richter knew a great deal about rats. He could use light to alter their activity pattern at will, or inject them with drugs to confuse their clocks. Skilled in making rats' clocks go, Richter decided to look for ways to make them stop.

Since the animals' adrenal glands regulated their activity cycle, he took these out. The rats slowed down a bit, but stayed regular. The pituitary regulated the adrenals, so he took out the pituitaries. The daily rhythms persisted. Richter supposed that the clock must be somewhere in the nervous system, so he went after parts of the rats' brains. He deprived them of oxygen, shocked them, froze them, anesthetized them and injected toxins. He stressed them, requiring one to swim in a torrent of water for forty-eight hours and forcing others into fights by giving them electric shocks. In reply, Richter's rats kept right on ticking, displaying regular cycles.

Nothing worked except destroying parts of the hypothalamus, an area of the brain located behind the eyes. With this organ damaged, rats chose to eat every forty to sixty minutes and drink every few minutes, going without sleep and wandering around in a daze the rest of the time. Richter concluded that various clocks throughout the body controlled various functions.

Marking Time in the Menagerie

While Richter was monitoring, and significantly modifying, his rats, other researchers looked for clocks in other animals. They found cycles of activity in species as different as hamsters—which ran on their exercise wheels only at certain hours—and South Pacific sea worms—which swarmed on specific days, at certain times. One researcher even figured out the formula creatures use to keep their clocks running on time.

Patricia DeCoursey lived in Madison, Wisconsin, where the nearby woods offered a lively display of flying-squirrel aerobatics. The squirrels left their dens at dusk each evening, skimmed from branch to branch during their period of activity, then returned to the same home trees. Since the time of dusk changed each day, DeCoursey wondered whether the squirrels were responding to light or to some internal clock.

She collected squirrels and isolated them in a basement lab where the sun never shone. She gave them wheels to run on, but kept the lights off all the time. Like Darwin's plants before them, and like Beling's bees, the squirrels kept obeying daily cycles of rest and activity.

But DeCoursey had more accurate tools than Darwin or Beling, and after measuring her squirrels' cycles, she quickly saw that they were not always twenty-four hours long. Some raced on their wheels every twenty-three hours; others took twenty-four and a half hours to complete their daily activities. What really piqued DeCoursey's curiosity was that each squirrel was consistent. Twenty-three-hour squirrels, if kept in darkness, always omitted one hour from their days. If they kept this up for twenty-four days, they would lose one full day. Others whose cycles ran more slowly, using an extra half hour to complete a day, got ahead of the sun. After long enough they might gain a day.

Without the sun to reset their watches, they just obeyed those inner rhythms.

DeCoursey decided to fake out her squirrels' clocks. She took the approach others had used on plants—that pulse-of-light-in-the-dead-of-night trick. The squirrels responded by resetting their free-running cycles, but the dimension of this resetting depended on what time the light arrived.

If the lights flashed when a squirrel thought that the hour should be around dusk—time to take off on the evening's flight—it would alter its flight plan radically. The light signal told it that the sun must still be shining brightly, definitely the wrong time to start the air show. The squirrel hid out for several more hours; and in the days that followed, all its activities were shifted to new times based on this one signal of light.

If the signal came in the middle of the squirrel's subjective daytime, when it would not be active, light had no effect on its rhythms. Light signals arriving during the subjective night caused a small effect on the rhythm, and this also persisted in the days that followed. By resetting their clocks, DeCoursey changed what phase of day they thought it was—"phase-shifting" their rhythms. Today scientists are incorporating DeCoursey's discoveries into research that seeks to phase-shift our own bodies' rhythms.

In DeCoursey's time, information about the full range of inner clocks was arriving from studies of a variety of animals. Monkeys, bats and even camels displayed daily cycles in the rise and fall of their temperatures. Animals' temperatures were usually highest during the hours they were active, lowest during their resting times. Other internal features also went through daily cycles. The amount of bile one could measure in a rabbit's liver depended on what hour one took the measurement. The effectiveness of chemicals that transmitted nerve impulses varied according to both the time of day and season of the year. The impact of drugs, such as those used to treat heart disease or insomnia in humans, was quite different when those drugs were given to animals at different times.

If drugs used to treat human illnesses had different effects at different hours, it might be time to rethink the whole concept of homeostasis. But before that could happen, science needed to acknowledge that human beings might have cycles at all.

6

Beat the Clock

In 1936 American crowds filled movie theaters to enjoy Charlie Chaplin's latest film—*Modern Times*. They laughed until they cried. Chaplin's comedic parable, about a man at the mercy of an assembly line, tweaked both the funny bone and the tear duct because it was all true. One had better laugh. There was little time for tears.

We were three decades into the twentieth century, and time had accelerated with a vengeance. Industry, transportation and communication were collapsing days into hours and hours into minutes. Expensive factory equipment could not be allowed to languish overnight just because people preferred to sleep. Fires needed to be kept burning in smelters. Until that enterprising fellow Henry Ford introduced the eight-hour shift in 1914, twelve-hour work days were the standard, and many of those days were nights.

Night-shift workers began paying the price. By 1929 their risk of developing stomach ulcers was estimated to be eight times that of day workers. That estimate later declined, perhaps because only the strong-stomached survived on night-shift work, but employers too began to wonder if there were drawbacks to pressing production onward through the wee hours.

An unusually high number of errors and accidents occurred during the night shift, their number peaking between 2:00 and 4:00 A.M. In addition, efficiency declined.

Yet these were years when the national economy was expanding at a frenetic pace. New machines needed round-the-clock workers to mind them, and plentiful natural resources required tireless workers to mine them. With all this work to do, the first natural resource being laid waste was sleep.

Yet sleep seemed unnecessary when so much was possible. Trains were racing across the continent. Airplanes could leap the oceans. Why, if fatigue were not a factor, a pilot might even fly all the way around the world. Or so thought the aviator Wiley Post.

A Race Against Sleep

In the early 1930s Wiley Post had a single-prop twenty-seven-foot plane and a dream big enough to honor his native state, Texas. Post determined that if he put himself into training for months—learning to sleep on an irregular schedule and developing acute concentration no matter what the conditions—he might make it around the globe in eight days. On June 23, 1931, he took off.

When he landed on July 1, having traversed 15,500 miles, Post captured success. He also had the first, and possibly the worst, case of jet lag on record.

At one airport, he stopped for fuel, but forgot the fill-up and just took off. He had to turn around and go back. At another, a newspaper reporter attempted an interview, but Post fell asleep in the middle of answering a question. When he finally did return to earth, Post got a ride in an automobile, but there seemed no escape. Even as he rode, he still felt as if he were flying.

If he'd gone on without sleep for too long, Post might have begun hallucinating. Studies were beginning to show that skipping sleep, whether to set a world record or punch a time clock, could have dangerous consequences. Yet even two decades later, by which time international flights were carrying more than two million passengers a year, the lesson of Wiley Post's adventure remained to be learned.

It was officially recognized in 1953, the year that John Foster Dulles, secretary of state, flew to Egypt to negotiate the treaty for the Aswan Dam. Dulles skimmed across seven time zones, landed and soon went into meetings. For Dulles, the meetings occurred at a biological hour when he should have been nose-to-nose with nothing tougher than his pillow. Dulles lost the deal to the Soviet Union. He had blown it, and Dulles knew why. Shortly after he returned home, he warned fellow diplomatic workers that jet lag was real. It was wiser to sleep first and answer questions later.

The reason for this was becoming clear from research done the very year that Dulles sleepwalked through Egypt. Scientists were finding that sleep was neither a luxury nor a constant state. No seamless span of unconsciousness, sleep was emerging, under careful study, as an inexorable daily cycle which, in its turn, contained precisely choreographed smaller cycles.

The Timely Cycle of Sleep

By the 1950s electroencephalographic recordings were revealing that human sleep displayed four separate stages, each characterized by a recognizable electrical signature. These stages repeated themselves every ninety minutes, and the most distinctive among them showed not only unique brain-wave tracings but also rapid movements of the sleeper's eyes behind closed lids.

This sleep stage got its name—REM—from those rapid eye movements. It also found an explanation. When sleepers' eyes were moving and their brain waves were revealing REM sleep, waking them interrupted dreams they could remember and report. When their eyes were unmoving and their brain waves showed a different sleep stage, they would awaken to report dreamless sleep. During REM sleep, the sleeper was dreaming.

The stages of REM and non-REM sleep alternated in a predictable rhythm throughout the night. Volunteers could be awakened every time they went into REM sleep, but when they slept again, the pattern on their EEGs changed. They spent extra time in REM sleep, presumably catching up. If researchers' efforts prevented them from getting their nightly allotment of REM for too long, some showed signs of emotional disturbance. Being awakened during other sleep stages caused no similar problems.

It began to look as if, on some level, our psychological well-being might be tied to our sleep rhythms. Yet sleep was not the only mental state yielding clues about how cycles affected our mental health. Manipulating waking states, too, could result in strange psychological consequences. Predictably for a history so liberally salted with serendipity, one of the first clues to this turned up quite by accident.

Frozen in Time

Like Columbus before him, the explorer Frederick A. Cook was not a man in the habit of finding what he went to look for. After numerous attempts, Cook claimed to have reached the North Pole. Others thought differently, and Robert E. Peary won that honor in 1909. Yet in his searches, Cook was among the first to record clear-cut sightings of a psychological malady that, nearly a century later, is still not fully understood.

During arctic winters, the sun neither rises nor sets, making it difficult to keep track of days. Like Bünning's bean plants and DeCoursey's squirrels, Cook and his team lived in fairly constant darkness, sleeping and working in what must have seemed like endless night.

Cook still managed to keep a diary, although not always accurately. Three days are missing from the records he kept of the winter of 1909. His team kept track of time by checking the phases of the moon, but every so often one of the men became convinced that they had lost a day. If that man was right, as the missing three diary entries suggest he was, Cook and his team may have been free-running like DeCoursey's squirrels.

They may also have felt like animals trapped in dark basement cages. As one of Cook's notations from northern latitudes acknowledged, "The curtain of blackness which has fallen over the outer world of icy desolation has also descended upon the inner world of our souls." His men sat around, sad and dejected, "lost in dreams of melancholy from which, now and then, one arouses with an empty attempt at enthusiasm. . . . All efforts to infuse bright hopes fail."

Other adventurers, following in Cook's and Peary's footsteps, also found their souls stained by the dark melancholy of arctic winters. Meanwhile, at lower latitudes, physicians were observing that some patients with mental disorders also suffered more during winter's dark days.

One doctor noted that, among mental patients, moodiness often arrived in autumn and cleared up in the spring. Another found that even among apparently mentally healthy individuals, moods followed regular cycles. When he observed industrial workers, he saw that their moods swung from happy to sad, and back again, on four- to six-week cycles. The changes were so subtle that his subjects never noticed them, but the evidence was clear from their responses to the doctor's interview questions.

A Clock for Every Purpose

By the late 1950s it began to look as though the human body was filled to the rafters with physiological clocks. Their cycles determined everything from when a single cell was born to the moment the body itself might die.

One study examined healthy tissue left over from surgeries and found that cells were most likely to divide in the late afternoon and early evening. Studies of diseased tissues showed that tumor cells followed their own distinct schedule of proliferation. Endocrinologists monitored the output of the adrenals and concluded that these glands secreted hormones on twenty-four-hour schedules. In human urine the concentration of chemicals also varied according to the hour.

Research on the treatment of ulcers revealed that gastric acids entered the stomach in greater or lesser amounts depending on the time of day. Even stomach contractions, those signals of hunger, came in a pattern reminiscent of the cycles of sleep—every ninety minutes. As for the end of life, if it was to come by a heart attack, death was likeliest shortly after awakening. A study of more than twelve hundred patients showed that their hearts most often lost life's rhythm at between eight and ten in the morning.

With so many reports flowing in, to so many research journals, it was inevitable that someone would put the pieces together. Scientists read each others' papers and fired off studies and letters. A threshold was reached. One researcher suggested a symposium; several formed a committee. In the summer of 1960, 150 detectives who had been pondering the clues about biological cycles finally came together to talk.

A Matter of Timing

Every so often science experiences a phenomenon called a paradigm shift. Such a shift takes the usual way of looking at things and turns it on its side, on its head, or thoroughly inside out. In Cold Springs Harbor, Long Island, a paradigm shift began in June of 1960.

Paradigm shifts occur because science, despite its best intentions, has a way of getting stuck in the past. A certain theory—for example, that the sun revolves around the earth—nicely fits what everyone can see. The sun comes up and goes down. That's hard to miss. The earth itself might be spinning, but people can feel it when they're spinning, and no one does. The theory looks true, so thinkers begin adding ideas onto it. They try to measure just how fast the sun is circling the earth, for example, or they look for the original push that got the sun moving.

Most of these secondary ideas fit the evidence pretty well. Occasionally one must make excuses for what should be a perfectly good idea—based on a moving sun—because the results don't quite support the main theory. One blames faulty observation, inaccurate instruments, the weather or, if all else fails, magic. Yet gradually a disturbing trend emerges. An increasing amount of evidence just won't fit the underlying supposition that the sun moves. Exceptions prove the rule, but soon these exceptions are numerous enough to overshadow the rule.

That's when science is ripe for a Copernicus, a Galileo, Newton or Einstein. Someone points out that the exceptions actually suggest a brand-new rule. If the earth revolved around the sun, for example, all those problems measuring the speed of the sun's movement or how it first got moving would go away. Yet old rules don't die easily.

Brickbats fly and so do insults. Warring camps found clubs, or else pick them up and start hitting each other. Occasionally someone gets banished or burned at the stake. A paradigm shift can be dangerous, but it also delivers a payoff. Gradually observers begin testing the new explanation, and it becomes accepted as true. Our image of the world—our paradigm—has shifted.

For the present, the theory that the earth revolves around the sun is expected to hold. That is, unless too many exceptions emerge to undermine it. Stay tuned.

Paradigm shifts can start suddenly, but decades may pass before their full impact is felt. Cold Springs Harbor was the site of one such jolt, and we are still picking up aftershocks.

Everyone who was anyone in the clock-watching world attended. Max Renner, who had confirmed that bees had clocks, came. So did Janet Harker, who had suggested the location of cockroaches' clocks. They encountered Colin Pittendrigh, who had experimented with everything from fruit flies to mammals. Another participant, Jürgen Aschoff, was at work formulating a set of rules for how light and darkness affected animals. He no doubt encountered Franz Halberg, who coined the term *chronobiology*. They were joined by researchers studying everything from agriculture to zoology, from cancer to cosmic rays.

Scientists came from such countries as Italy, Germany, Japan and Sudan, as well as from the United States. In ten days they shared what they had learned about rhythms in nature, and by the time they departed, few among them would question that life on every level was governed by living clocks. They returned to their laboratories to explore the implications of the new paradigm. Joined by others, they've been at it now for slightly over three decades. Those mystery clues that began arriving more than two thousand years ago are finally fitting into place.

We know that we have clocks. We know a lot about how they work and why they stop. We are even closing in on one or two locations in our bodies that may act as master clocks—setting the pace for how our inner lives are ruled by time.

III

What We Know

7

A Tour Inside the Clockworks

T he human-made clocks we watch every day show seconds, minutes and hours. Inside our bodies, time is divided quite differently. Cycles may last mere milliseconds or require hours to complete. We might expect such a wide variety of rhythms to display a completely random range of periods—3.7 seconds perhaps, 18.6 minutes, fourteen hours, and every stop in between—but they do not. Instead, our inner clocks prefer certain periods for their cycles.

They divide time into seconds, or portions thereof, minutes, ninety-minute blocks, and eight- and twelve-hour lengths. Many of these preferred periods add up nicely to the length of a day. It's as if evolution set itself the problem, "How many ways can I divide a twenty-four-hour day?" We are the answer.

A Hierarchy of Clocks

Chronobiologists classify our clocks into three basic types, and have built rabbit warrens of subtypes descending beneath these. In general terms, our biological clocks are either *ultradian*, *circadian* or *infradian*.

Ultradian clocks keep time in units shorter than a day. Their rhythms may be only milliseconds long—like the cycle in which neurons fire— or a few minutes or hours long—like the ninety-minute sleep cycle. Chronobiologists suppose some of the smaller clocks might drive the larger ones, or vice versa, but they're not sure.

The second type, circadian clocks, measure time in periods about twenty-four hours long. These range from the obvious pattern of sleeping and waking to subtler daily rhythms of when we're most sensitive to pain, how well we estimate the passage of time, or when our bodies are most affected by alcohol. So far the greatest number of known clocks are the circadian ones.

The third type, infradian clocks, run on periods longer than twenty-four hours. The monthly rhythm of women's menstrual cycles provides the clearest example of these, and research is now suggesting that men also have monthly cycles. In addition, humans have cycles that take place over several days, several months, and the year. Yearly cycles are well recognized in the animal kingdom, but humans, too, are more likely to gain or lose weight, bear offspring or feel sleepy depending on the time of year.

The longest cycle of all, of course, is a lifetime. As far as science knows, we only live once, so the period of this cycle is one per customer. Yet within this lifetime, our inner cycles shift and change. A normal temperature rhythm for a twenty-year-old might be abnormal for an eighty-year-old, and if a thirty-year-old actually slept like a baby, something would be seriously wrong.

Ultradians, circadians and infradians measure out our lifetimes in intricate, innerconnected patterns. When scientists isolate us and listen very closely, they can actually hear us ticking.

8

What's Inside a Day?

Among ultradian rhythms, our swiftest-running clocks do the work of the nervous system. They mark time with the one hundred billion neurons of our brains. These clocks are literally as quick as thought, sending intangible signals racing along nerve fibers. When we're resting, an electroencephalograph shows the brain's electrical rhythms averaging from eight to twelve cycles per second. During the morning, our brain waves cycle at the slow end of this range. They speed up when darkness falls and cycle most quickly when we're asleep.

Slightly slower clocks oversee the delivery of tangible items— blood, oxygen and food. We can hear and feel the ticking of these clocks. Our hearts beat about once every second, and on average we take and release one breath for every four heartbeats, about fifteen times a minute. We blink about every two and a half seconds.

Meanwhile, the muscles of our intestines contract about every minute, with our stomach's muscles sending along waves of peristalsis, or involuntary contractions, every three minutes or so. Our skeletal muscles display four- to six-minute cycles in strength. Even isolated sections of muscle or heart—removed from animals—show spontaneous rhythms.

Such short-cycle, ultradian clocks must be remarkably small, perhaps even as small as a single gene. Quick-cycling rhythms appear to be built into all living creatures at the genetic level. We know this because biologists can now transplant genes to construct sixty-second living clocks.

Using this technique, they're putting together a plant with one gene for photoluminescence—causing it to glow in the dark—and another that governs timing. If they succeed, they'll build a plant that lights up on a regular schedule. Perhaps one day we'll someday enjoy our flower gardens at night, too, when they give light shows.

The Sixteen-Hour Day

Above the minute level, our biological clocks divide time into larger units, but they construct an entirely different day from the kind we're used to. Apparently our bodies have two types of "hours"—the usual sixty-minute one and another that takes a half hour longer. Thus, for some hormonal rhythms, a biological day may be made up of twenty-four cycles of sixty minutes each. For our brains, however, a day may include sixteen cycles of ninety minutes apiece.

Cycles displaying a sixteen-hour "day" first became evident when scientists began studying sleep. They found that in ninety minutes we go through four distinct stages, as well as the state of rapid eye movement, or REM. Sleep represents a constant state of change.

For a few minutes after we settle down, our brains drift, their electrical waves cycling three to seven times per second. This pattern, measured by an electroencephalograph, looks a lot like the pattern our brain waves make when we're awake. Gradually our breathing and heartbeats slow and become regular. This is stage 1, a transition, where we spend only about five percent of the night. Unless we're awakened, it does not occur again during a given night. When we're drifting like this, we can be awakened easily, and those who do awaken report dreamlike hallucinations.

About ten minutes later we become much more difficult to arouse. If we do wake up, our sleeping thoughts are fragmentary and hard to recall. We have moved into stage 2, with the brain's electrical activity cycling twelve to fifteen times per second. Characteristic spikes and patterns appear on our EEGs. Our blood pressure declines, and our brains are using less of the oxygen the bloodstream delivers.

Meanwhile, from near the base of the brain, the pituitary gland delivers a large dose of growth hormone to the bloodstream. Most of our blood goes to our muscles, helping repair the wear and tear of the day. We can move our large muscles during this time, but we'll be unable to do this in a deeper stage. This stage of sleep, which we'll return to regularly through the night, takes up about forty to fifty percent of our nights.

Stage 3, where we spend five to seven percent of the night, acts primarily as a transition to the next stage. Along with stage 4, it is known as slow-wave sleep, or deep sleep. Our EEGs scrawl huge, slow ripples as slowly as once per second. Our blood pressure, pulse and temperature decline. Our muscles go limp. If someone wakes us, we feel disoriented.

Sleep's most dramatic stage arrives next. Rapid eye movement sleep is the dark cave where we find dreams. For most of us, this stage begins abruptly about ninety minutes after we drift off. We're fairly easy to awaken when we're in REM, even though we're truly in another world.

On a sleeper's EEG the pattern begins to look saw-toothed, like a bad case of the jitters. If we've been snoring, we fall silent. Our blood pressure, respiration and pulse become irregular and may fluctuate wildly. As much as a quarter of the oxygen in our blood is used by the brain. There nerve cells have relaxed our large muscles so completely that we are nearly paralyzed. We may make small, jerky movements—our faces and limbs twitching—but we're mostly still.

It's a good thing, too, because we're very busy. Behind closed lids, our eyeballs flick from side to side, up and down, all around. The visual portions of our brains are crackling with electrical activity. We're seeing our dreams, but if we acted on what we saw, we might hurt ourselves.

Other parts of our bodies also show signs of REM. More blood flows to our genitals. Men, whether they are infants, mature or aged, and even those who are impotent, may become aroused. Women, too, often show signs of arousal, and a woman's uterus may contract. These responses have nothing to do with the particular dream going on—a fact that probably disappointed researchers who learned it by waking subjects and asking for details.

The first REM cycle of the night lasts about five minutes. Then the brain's electrical activity slows again, and we return to stage 2. We move through stages 3 and 4, coming back to REM ninety minutes later.

The second time around, REM may last ten minutes. On our third visit, another ninety minutes later, REM can extend for fifteen minutes or longer. As REM cycles lengthen, our non-REM stages— particularly stages 3 and 4—grow shorter. By the early morning, after we've gone through this cycle four or five times, we may stay in REM for a full hour and only briefly visit stages 3 and 4 before returning to REM. We spend as much as a quarter of our sleeping time in dreams, and the longest dreams come near dawn.

No one yet completely understands why we go through these ninety-minute cycles. It may be that non-REM sleep uses growth hormone to restore our bodies, while REM sleep keeps our minds emotionally balanced. Several studies indicate that our amount of REM sleep increases when we're placed in new situations or required to learn new skills. Yet even that explanation does not account for why we would also experience ninety-minute cycles while we're awake.

The Ninety-Minute Hour

At first, scientists thought this ninety-minute pattern only occurred during sleep, but eventually they noticed that our brains continue living on sixteen-hour days even when we're awake. Test subjects kept in bed all day—in an experiment for which it must have been easy to get volunteers—tended to move around more about every hour and a half. In another experiment, subjects were kept in isolation, but with free access to food and coffee. They headed for the refrigerator or the coffeepot about every ninety minutes.

Another research project, testing our sensitivity to pain, probably had greater difficulty getting volunteers. Subjects reacted to the application of cold to their front teeth and showed that pain thresholds varied on approximately ninety-minute schedules.

We show ninety-minute cycles in the secretion of some hormones, as well as in rise and fall of our appetite and attention span. This may be because the brain itself changes what it's doing every hour and a half.

In the late 1970s a couple of researchers set out to test whether our thinking abilities also display ninety-minute cycles. They knew that the brain's two hemispheres excel at different types of thinking—with the left side supporting our verbal abilities and the right side looking after our visual ones. They decided to test these abilities over an eight-hour period. Every fifteen minutes they had their subjects quickly identify pairs of letters and match up patterns of dots. As expected, the testees did better with the letters at certain times, better with the dots at others. One skill improved as the other declined, and the best performances alternated in cycles that were—not surprisingly—about ninety minutes long.

Thus it looks as if our bodies prefer hours that last an extra thirty minutes. Since 90 divides neatly into 1,440—the number of minutes in twenty-four hours—these cycles stay pretty much in line with our usual days.

It may be that this rhythm represents a basic internal cycle of rest and activity, such as those that animals display. Mice vary their activities every twenty to thirty minutes, and cats do so every thirty minutes. Monkeys keep time in forty-five-minute hours. These cycles tend to be briefer in lighter-weight species, longer in heavier ones. We heavy humans may have a ninety-minute basic rest activity cycle, but the real heavies are elephants. Their cycles of rest and activity are 120 minutes long.

On Hours and Off Hours

While our ninety-minute cycles go on, awake or asleep, inner systems also keep time in even longer units. Beyond the ninety-minute level, our days reveal rhythms several hours long.

The strange-but-true prize for inner clocks has to go to the cycle of breathing. On average, we breathe through one nostril for three hours, with the tissue in the other nostril slightly engorged, then we switch. This change is usually too subtle to notice, unless, of course, we have a cold.

Nor are we likely to notice the cycles of hormones arriving in our bloodstream, released on set schedules that range from two to four hours in length. As for sleepiness during the day, we feel most like napping precisely twelve hours after the middle of our last sleep. Those 3 P.M. yawns mirror our 3 A.M. snores. Apparently our bodies believe in siestas.

As the milliseconds, minutes and hours add up, our ultradian rhythms accumulate into what seems the most natural unit of biological time—the day.

9

As the World Turns

If a hypochondriac had nothing better to do and arranged to have two medical workups—one at 8:00 A.M. and the other at 5:00 P.M.—he might persuade himself that a space alien had taken over his body that day. His blood pressure, pulse, metabolism, brain waves, respiration, urine, temperature—in short, just about everything except eye color and shoe size—would have changed. Some readings would be as much as three hundred times what they were hours earlier. The hypochondriac, like all of us, would be radically different at noon from what he was at midnight, but there's no need to credit space aliens with the alterations.

Daily Ups and Downs

On the outside we look pretty much the same whether it's noon or midnight. On the inside we're as different, day to night, as day is different from night. These changes go on regardless of what we happen to be doing or how we expect to feel.

In late afternoon our blood pressure will rise even if we're plugged into biofeedback and meditating to lower it. Our skill at driving will be poorest just before dawn even if we're stone-cold sober, chewing gum, fresh from a nap, and playing the radio. We'll find it easier to remove the tight lid from a jar at around two in the afternoon—when our grip

58

strength peaks—than at dinnertime, no matter how thoroughly we curse it. Our biological rhythms carve peaks and valleys into our days, altering who we are and what we can accomplish.

If all the volunteers tested to develop these norms were represented in one composite man, he would be a statistical chimera. Norm the Average would be so typical that his wife could easily misplace him at shopping malls, and his dog would often follow similar men home. Norm would represent forty-nine percent of us, while Norma the Average, his wife, would represent women, the other fifty-one percent. Their dog, of course, is Every Dog. Looking at one of their days reveals our basic circadian cycles.

The Clock Strikes Twelve

Around midnight on a weekday, Norm the Average would be sleeping like a baby. He hit the pillow two hours ago, so his REM and non-REM cycles now alternate gracefully, causing him to snore intermittently, then pause for a dream. Mrs. Average, snoozing beside him, takes turns snoring when Norm is silent.

If there's a fountain of youth in this world, its elixir flows in Norm's veins right now. During the first few hours of sleep, Norm's body releases about half of his daily supply of growth hormone. In animals, this hormone has been shown to reduce cholesterol and to cut risks of cancer, diabetes and stroke. In fact, the proportionately higher level of this miracle substance flowing in Norma the Typical's veins—women secrete more growth hormone than men do—may be part of the reason that women often live longer. Norma will lose her advantage as she ages, however. As both men and women grow older, their levels of this rejuvenating potion decline.

About an hour after midnight, Norm's skin cells are dividing at their fastest rate for the day, but not much else is happening. The amount of DNA being made in his body is down to one tenth of what it will be at other hours. An unusually low number of cancer-killing cells circulates in his bloodstream. His blood pressure is moving toward its low point, and so is his temperature. Meanwhile, Norm's mental ability is careening downhill.

Norm and Norma's brains are awash with a chemical called melatonin, a substance whose job it is to make them sleepy. No matter what we're doing during the still, small hours of the night, our melatonin flows. Its concentration peaks around 2 A.M., four to six times higher than daylight levels. If we're awake, our ability to do even simple tasks is swamped by this high tide.

This is probably why many industrial and traffic accidents occur during the wee hours. One study found that single-vehicle accidents increased sixteenfold in the early morning hours, even though most of the drivers got eight hours sleep the night before. With our melatonin flowing, our brains may cycle us though REM and non-REM episodes against our will. We may lapse into microsleeps and wink out for two to three seconds, taking a wrong turn, turning the wrong valve or pushing the wrong button. The deadly chemical release at Bophal occurred shortly after midnight. Chernobyl's nuclear reactor exploded at 1:23 A.M., and the accident at Three Mile Island began at 4:00 A.M.

Another reason that Norm and Norma are least efficient now may be due to the decline in body temperature. Between 3:00 and 4:00 A.M. our temperature, which can reach a normal high of 99 degrees during the day, may sink as low as 97. Our abilities to think clearly and react quickly plunge along with it. Women's temperatures are generally higher than men's, so Norma's body is slightly warmer than her husband's, but she follows the same daily cycle.

In one odd observation, scientists have found that the skin's temperature varies depending on the side of the body. During daylight we are slightly warmer on our right sides; during sleep, slightly warmer on the left. Average Norm rolls over, turning a warm shoulder to his wife.

Norm and Norma take turns snoring, but their capacity to do so is diminishing. Our bronchial passages tighten up during the night, so right now Norm and Norma are most susceptible to breathing difficulties. In addition, in case either has been putting off a visit to the dentist, these are the hours when a toothache may start throbbing. Our sensitivity to tooth pain—as determined by experiments one shudders to contemplate—is lowest between 3:00 and 8:00 A.M.

Single-digit hours pass; morning approaches. The couple's heart rates begin to rise, and so does their blood pressure. Meanwhile, adrenaline is entering their bloodstreams to begin preparing them for the day.

Four to five hours before these two awaken, an important substance begins filtering into their bodies. This is cortisol, the powerful hormone that gives us our spark, our energy. A quarter of our daily total is pumped out by our adrenal glands in the few hours before we awaken.

Even if Norm had stayed up all night, perhaps disturbed by Every Dog's barking, his cortisol would still follow this pattern.

Since Norm and Norma are so average, they will sleep precisely 7.8 hours. Only one in one hundred of us is a "short sleeper," needing four and half hours per night. Another one in one hundred, the "long sleeper," needs ten and a half hours.

Women are generally shorter sleepers than men, so Norma awakens. This awakening is spontaneous, thanks to an inner alarm keyed to her ninety-minute sleep cycles. She checks the time and shakes her husband's shoulder. It is 6:00 A.M. Norm's eyes open. The most important chronobiological event of his day occurs.

The Body's Wake-up Call

Light strikes Norm's eyes. Instantly a complex biological process begins. His eyes absorb the light, and visual pigments set off a cascade of chemical reactions that amplify the signal and, like an electrical transformer, convert it into a code that his brain can read. There, in an unpronounceable locale known as the suprachiasmatic nucleus—handily abbreviated to SCN—reactions begin occurring. Eventually they will update many of Norm's inner clocks, letting all the parts of his body know when "today" officially began.

So far as chronobiologists know, the SCN is one of at least two "master clocks" that keep Norm's many inner clocks running accurately. If Norm and Norma were being kept in isolation down in a cave, they might not get this signal, and many of their body clocks could begin to free-run.

Fortunately, Norm and Norma are above ground. They crawl from bed and begin do-si-do-ing in and out of the shower, at the bathroom sink and around the coffeepot. Norm plants a kiss on his wife's head, perhaps because—like most men—his testosterone level is higher in the morning than it will be for the remainder of the day. No matter how Norma responds, Norm's pulse is kicking up. It is as much as fifty percent higher now than it was in the middle of the night. Meanwhile, changes are also taking place in Norm's blood. If he cuts himself shaving, he will bleed less than he would later on. Platelets simply stick together better in the morning.

Norm and Norma skip breakfast, and that's a shame. Both of them could stand to shed a few pounds. They should eat now.

Our bodies handle food differently depending on the time of day. In experiments that limited participants to a single meal each day, those who ate only breakfast lost weight, but some who ate only dinner gained it. Breakfast appears to be the meal our bodies convert most efficiently to energy rather than fat.

Regrettably, our average couple settles for only caffeine and vitamins. At least Norm also takes an aspirin. His doctor has warned him about heart disease.

It is 7:00 A.M., a good time for Norm to be thinking about his heart. The waking hours put increased demand on it, with blood pressure and heart rate rising, plus those stickier platelets. Most heart attacks occur around nine in the morning. As for that aspirin, between 7:00 and 8:00 A.M. the smallest dose of this drug has its greatest effect. Because of his kidney's schedule for processing various substances, Norm's aspirin will remain in his body for twenty-three hours. It would only stay for only seventeen if he'd taken it with dinner.

Interestingly enough, if Norm had a drinking problem this would also be a good time for him to indulge. A man becomes the least inebriated from a drink taken at 7:00 A.M., and the most from one taken at eleven o'clock at night. As for Norma, her body metabolizes alcohol most quickly at three in the afternoon. Neither Norm nor Norma happens to have a drinking problem, but Every Dog does. Every Dog emptied his water dish last night, and now he scratches at the door for his walk.

Energetic Mornings

Even those who have been deprived of sleep feel a surge of alertness in the morning, and Norm has slept well. The level of his liquid get-up-and-go, that cortisol, is peaking. He grabs the leash and heads out for a jog with Every Dog. Meanwhile, Norma leaves for work.

Twenty minutes later she arrives at the bank where she works as an administrative assistant. Norma is feeling tip-top, in more ways than she knows, and she begins putting in a highly productive morning.

As she works, the divisions among her skin cells are becoming fewer, but inside her bone marrow the cell duplication that will fuel her immunities is starting to rise. This will put more cancer-killing cells into her bloodstream. Elsewhere in her body, different organs, and even different types of cells within organs, will duplicate themselves at set times throughout the day. At some hours, cells may hardly be dividing at all. At others, ten times as much DNA is being made.

As the morning hours pass, Norma's mental abilities improve. The level of melatonin in her bloodstream drops, and her adrenal hormones make her increasingly alert. Even if she were commanded to fall asleep, it would be hardest for Norma to do so now. We are most resistant to sleepiness at two points in the day—five hours after our temperatures hit their nighttime lows and again eight hours before.

Norma's mornings are filled with minutiae, and she's good at that. Her concentration and her short-term memory are best between 9:00 and 11:00 A.M., and so is her ability to estimate what time it is without glancing at her watch. Norma is the quintessential one-minute manager.

Meanwhile, on the other side of town, Norma's husband has arrived at the Average Products Factory. Norm works here as a supervisor. During the morning hours, both he and his workers are experiencing similar benefits from their circadian rhythms.

Everyone's level of alertness is rising, and on the factory floor, according to tests at similar companies, the highest number of Average Products is now being produced. The workers' alertness hits its peak around noon, then declines for a couple of hours.

Norm is fortunate in that the Average Products Company cares about its employees' health. After a late lunch he heads down to the basement exercise room for a workout. He tosses off ten extra push-ups as if they were finger exercises and feels delight at his improving fitness. Norm's body has, however, worked a flattering deception on him.

Everyone's strength peaks in the afternoon. That's when we're best at doing sit-ups, push-ups and such, and when our lungs are operating at peak capacity. For example, nearly all the track-and-field records set in the last century came in events scheduled in the afternoon. Even those pedaling stationary bicycles do better late in the day, as much as seventy percent better.

Norm's exercise session will pay off nonetheless. Vigorous activity alters the rhythms of sleep, so Norm will experience more slow-wave sleep tonight. That means more growth hormone will enter his bloodstream. Unfortunately, as he rides the elevator back to the shop floor,

Norm begins to yawn. He blames the workout. He wishes he could settle onto the couch in the employees' lounge for twenty minutes or forty winks, whichever comes first.

The Afternoon Slump

In craving that nap, Norm is being incorrigibly average, as usual. Our alertness hits a temporary trough around midafternoon, twelve hours after our temperature hits bottom during sleep. Like the average American, Norm takes one or two naps a week.

One quarter of us never nap at all, but another third make up the difference by nodding off four or more times weekly. The average nap—the kind Norm really wants—lasts an hour and a half. Most of that sleep is in the deepest stage, with little time spent dreaming.

If Norm had not gotten enough sleep the night before, a nap could improve his alertness and give him energy, although the main benefit would be to his mood rather than his true mental ability. After napping, people do feel refreshed and perform better on tests of attention and complicated decision-making. Most people say they feel happiest at midafternoon, although they may be the ones doing all the napping. The rest of us experience a drop in efficiency, no matter what we're doing. Our work performance falls off, and traffic accidents show a peak parallel to the one that occurred twelve hours earlier.

Riding in the elevator upward, Norm thinks about that nap. It would probably do his heart good, he reasons. Tempting as that theory may be, nodding off would actually have little effect on factors influencing his heart.

The level of cholesterol in Norm's bloodstream is now approaching its peak, regardless of whether he ate oatmeal-bran-on-whole-wheat for lunch and did two hundred extra sit-ups. While Norm's heart rate is somewhat slower than it was this morning, its output and stroke volume are highest around midafternoon. This circadian pattern in heart rate persists even in those who have had heart transplants—a procedure that Norm fervently hopes to avoid.

Another measure of Norm's cardiovascular fitness is also changing. Inside our bodies, many features that affect blood pressure—the size of arteries, the force of the heart's contraction, adrenal hormones and even the amount of blood being pumped—follow strict circadian schedules. Depending on the time of day, our blood pressure can fluctuate by as much as thirty percent, and it reaches its peak in the late afternoon.

As he steps from the elevator, Norm remembers two things that jolt his blood pressure up another notch. Before he leaves today, he must crunch a complex series of numbers for tomorrow's production report. In addition, he must leave early for a dental appointment.

All hopes of a snooze dashed, Norm sprints the fifty yards to his office. The hour is approaching 4:00 P.M., and it's a good thing. Norm will need the circadian payoffs of this time of day.

His strength, flexibility and athletic skill—including his speed on that fifty-yard dash—are near their peaks. So is his quickness at addition, multiplication and even counting on his fingers. Norm works up a good mental sweat plugging a few last numbers into his report, then quickly memorizes the bottom line to recite it at tomorrow's production meeting. Since his long-term memory is improving as the day wanes, he'll probably remember all those random numbers.

Norm grabs his jacket and hurries the two blocks to his dentist's office. Thirty minutes later he is back on the sidewalk, numb on his left cheek but stunned at how little pain he felt during all that drilling. Even the shot hardly hurt.

Once again Norm has reaped the rewards of being incredibly average. If he'd gone to the dentist this morning, Norm would probably be feeling very sorry for himself right now. Yet during the afternoon his threshold for tooth pain is thirty percent higher than it was earlier. One experiment even found that lidocaine, injected to anesthetize those really sensitive front teeth, worked longest if given between 1:00 and 3:00 P.M.

Norm is relieved that the dental appointment went so well. Unfortunately, he had forgotten he had it scheduled when he made plans to meet his wife, Norma, for dinner tonight.

In the Realm of the Senses

Norm meets Norma at Chez Tres Ordinaire, their favorite entirely average French restaurant. Both love the cuisine here, although tonight only Norma will enjoy the delicate flavors. Norm's tongue feels like cotton batting.

Norm and Norma dine early, between 5:00 and 7:00 P.M. This is the time when, thanks to hormonal cycles, the average person's senses of taste and smell are at their sharpest. Norma raves over the trout almondine. Her husband picks at his coq au vin, his mood foul. M. de Median, the maitre d', seated them near the kitchen, and Norm's sense of hearing and sight are also at their most acute. He struggles to ignore the sight of dirty dishes going past and to converse with Norma over a riot of clattering cutlery and fluent French cursing. At least after dinner they have tickets for the symphony. Norm's ability to discriminate sounds has fallen slightly by 8:00 P.M., when the house lights go dim, but he is not disappointed in the performance.

Interesting things have been happening inside Norm's body, and in the bodies of those around him. His temperature, which has been rising all day, now is declining. It began going down around dusk and will drop a full degree before he goes to bed.

This temperature cycle is one of the most predictable of chrono-biological indicators. Many measures of performance mirror the temperature cycle, rising as it goes up, declining as it falls. Yet while the temperature rhythm is clearly endogenous, originating from inside the bodies of even those who are kept in isolation, it is also susceptible to alteration.

In one experiment, bedridden patients were fed through tubes on two different schedules. One group received its liquid nutrient during the day; the other group at night. The temperature cycles of patients fed during daylight followed the normal schedule, rising all day and falling overnight. Those who were fed throughout the night showed the reverse pattern, with their average temperatures peaking around 3:00 A.M. This suggests that food, like light, might be a powerful time giver for humans. It has long been known that food does this for animals, whose temperature cycles shift when the lab chow arrives at noon instead of midnight.

In addition, humans can apparently alter their temperature cycles if they really want to do so. That capacity helps make tonight's symphony performance outstanding.

A study of the circadian rhythms of members of British Broadcasting Corporation's orchestra turned up an unexpected finding. The musicians were a normal crowd, showing classic temperature cycles, except on performance days. On those dates, as opposed to days when they were merely rehearsing, the musicians somehow altered their temperature cycles so that their usual morning rise was lower and their primary peak came later—around the time of their evening performance. Thus, their mental and motor skills were probably better too. They put on a good show. Remarkably, this circadian shift showed up in the symphony's management and staff as well. Even for those who are not professional performers, early evening is the best time to practice or perform on a musical instrument.

Norm and Norma sink into their seats to take in their symphony's hottest concert. They sink down more heavily than usual, in fact, because their weight reaches a daily peak in the evening. Unaware of this, our average couple enjoys the performance of Banal's unexceptional *Musique Moyene*.

Melatonin is once again filtering into their bloodstreams, relaxing them. Norm yawns once or twice, since we become sleepy when our temperature is falling. His blood pressure and heart rate descend, too, and his alertness flags. It is nearly eleven when the final crescendo sounds, but Norm and Norma decide to stop for a drink on the way home. That's a big mistake.

Alcohol is among only a handful of chemicals that can reset our body's clocks, moving their hands backward or forward. In the course of a day, we experience as much as a twenty-five percent change in how rapidly our bodies metabolize this substance, and alcohol impairs our performance most in the evening.

Unfortunately for Norm, 11:00 P.M. is the hour when a single drink has its strongest effect. His wife orders lime and tonic, but Norm has three old-fashioneds. Becoming slightly intoxicated, he alters the circadian organization of his adrenal system for the next three days, raising his adrenaline to more than ten times its normal level. Wisely, Norma drives the couple home from the Usual Bar and Grill.

Felled by melatonin, inebriation and the fact that his level of cortisol has hit its daily low, Norm heads for bed. His wife putters around for a while. She calls Every Dog in from the yard and feeds him his Middling Meal. Norma brushes her teeth, having chosen the right time of day for this task too.

Along with the many inner shifts going on in the course of the day, our saliva also changes. During the night hours, we salivate least; and since saliva provides some protection against tooth decay, it's best to brush our teeth before bed. If we're going to have dry teeth, at least they'll be clean teeth.

Norma also puts on some skin cream, perhaps Liposome Typquical. Its fancy label does not specifically recommend using it overnight, but perhaps it should. Researchers testing a liposome face cream found that women's skin showed the best texture and most natural brilliance at ten in the morning, but that their cream was most effective if applied between ten and eleven at night.

Norma settles in beside her husband and begins to drift. Her last thought, before her brain waves change their pattern, is that by staying up late, they've cheated on how much sleep they'll get tonight. They're like twenty percent of all Americans in this, building up a sleep debt. They'll try to make up the difference by sleeping in late on the weekend.

Norm's snoring falls silent as he plunges into a dream. Norma begins snoring. Every Dog turns around three times and collapses with a sigh.

10

As Weeks and Months
Go By

T he workaday world where Norm and Norma live—where most of us live—follows a pattern of weeks and months as much as it does of days. Monday morning's rush hour carries us into a week that ends in T.G.I.F. On weekends we restore our bodies and spirits, then head out for another week. We collect paychecks, then pay bills, catch up on the latest in magazines and turn a calendar page once a month.

Because we live in modern times, we assume these schedules are imposed on our bodies from outside. Research into our inner rhythms reveals, however, that our time-clock world may have reinvented a system originally designed by nature. Not only do our bodies follow an approximately thirty-day moon cycle, they somehow know about the seven-day week as well.

All in a Week's Timing

When chronobiologists first noticed seven-day cycles in human blood, urine and immunities, they doubted they'd really found anything. After all, people did different things on different days of the week. Surely *circaseptan*, or seven-day, rhythms must occur because we organize our activities this way.

Just to be certain, they checked other species. Insects, including flies, revealed weekly rhythms. So did rodents, one-celled organisms and even mushrooms. None of these creatures worked forty-hour weeks. When humans were placed in isolation—out of touch with the weekly world—their bodies persisted in showing patterns lasting five to nine days. Eventually researchers concluded that some sort of seven-day plan must be built into our cellular machinery. The circaseptans were less pronounced than daily or monthly cycles, but they apparently formed part of our evolutionary heritage.

So far we don't know how circaseptans work, but numerous sightings confirm them. Our blood pressure and heart rate take about seven days to readjust after major schedule changes. One scientist tested his urine each day for fifteen years and found that its acidity and salt content varied about every seven days. Our temperatures follow seven-day patterns, and those who experience the pain of osteoarthritis or migraine headaches often find that their symptoms rise and fall every seven days.

The best-investigated circaseptan rhythms concern our immunities. The common cold, for example, will last a week no matter how much we spend at the drug store. Patients with pneumonia or malaria face the greatest danger around the seventh day of their illnesses, and the symptoms of chicken pox usually appear around two weeks after exposure. The immune cells vital for our resistance to infections and cancer—T-cells and B-cells—fluctuate in number on a seven-day schedule.

This immune cycle can make a difference if we undergo surgery. After surgery, for example, the amount of swelling that patients experience varies on a seven-day cycle, worsening on the seventh, fourteenth and twenty-second days. Doctors who perform kidney transplants know that the risk of rejection is highest one week after the procedure, and for a while danger zones continue to occur every seven days. Studies of animals have found similarly timed risks for heart and pancreas transplants.

In the debate about which came first—the social week or the biological one—biology appears to be winning. Those who speculate about such things suppose that nature must have divided the time it takes the moon to complete a full cycle—29.5 days—into four smaller chunks. That comes to 7.4 days per chunk. Society just rounded it off.

The Clock on the Face of the Moon

The 29.5-day moon cycle does exert a powerful influence on plants and animals. Marine animals keep tabs on the tides to know when to mate and find food. Night animals survive or perish depending on whether they hunt or stay hidden on moonlit nights. Daylight animals, too, exhibit cycles of activity and reproduction lasting about 29.5 days, and so do we.

Women's menstrual cycles provide the most familiar example. The word *menses* originally meant "lunar month," and during their years of fertility, most women menstruate every 29.5 days. This is an average, of course, since normal cycle lengths vary by as much as twenty-five percent. They may also be disrupted when we change our schedules.

Airline stewardesses, who lose and gain time when they fly east or west, have a higher likelihood of experiencing irregular menstrual cycles, usually longer ones. Women who work nights or on rotating shifts are more likely to have irregular cycles, or uncomfortable ones. If their sleep-wake patterns are consistently out of phase with the day-night world, they have more difficulty becoming pregnant, and when they do conceive, they experience more problems with pregnancy and more stillbirths.

Because biological rhythms intermesh, the menstrual cycle also predicts other changes taking place inside a woman's body. The most pronounced of these is the pattern of change in the immune system.

On the first day of menstruation, a woman is most susceptible to allergic reactions, thirty percent more so than at midmonth. This is also the likeliest time for tonsillitis, hives, herpes sores, ulcer attacks, the symptoms of glacoma and the risk of diabetic coma. Headaches are also more common at this time and during the few days before menstruation. Sixty-six percent of women's headaches occur on menstrual and premenstrual days, more than twice the number that would be expected if the rate were random. Asthma attacks display a similar pattern, with seventy-five percent of them occurring either during or shortly before menstruation.

At midmonth, around the time of ovulation, women face greater risks of catching colds or the flu. One study found that seventy-seven percent of the women exposed to cold viruses during midcycle caught the cold; only twenty-nine percent of those not at midcycle started sneezing. In general, women are most vulnerable to bacterial infection during this time, and the cells of their bodies may be at greater risk for genetic damage.

The danger from viral infections, such as hepatitis and viral pneumonia, dominates the latter half of the menstrual month. During this time, women experience more skin problems, including acne and eczema. For those who have epilepsy, the risk of a seizure increases significantly during the three or four days before the menstruation begins; and on the first day of bleeding, this risk can be as twice as high as it will be later in the month.

These changing levels of susceptibility may even affect a woman's risk of death. When researchers reviewed deaths of more than one hundred women, aged eighteen to forty-six, they found that only thirteen of them died during the first half of their cycle, before ovulation. The remainder died during the latter half, with sixty of the deaths occurring in a single week, between days seventeen and twenty-three.

Men's Months

The fact that women experience many changes during menstrual cycles has often mystified and muddled society's beliefs about them, and about their abilities. The vast majority of research attempting to relate the menstrual month to women's intellectual performance has discovered no connection. Meanwhile, history has shown an incurious reluctance to ask whether men's bodies also display monthly cycles. When researchers did look, they found surprises.

One seventeenth-century Italian scientist weighed men each day and learned that their weight fluctuated by an average of one to two pounds over the course of each month. More recently a Danish endocrinologist performed analyses of the hormones in men's urine and found a thirty-day rhythm of change. Men's beards grow more or less depending on where they are in a monthly cycle, and their temperatures and thresholds for pain also vary. When a scientist kept track of his lung capacity and grip strength daily for twenty years, he found monthly cycles in both.

One male medical problem definitely appears related to the moon's month. Many men experience benign enlargements of their prostate glands, a condition that can make urination difficult and eventually require surgery. Those with this problem are more likely to experience symptoms during the new moon.

For all of us, monthly cycles bring strengths and weaknesses. Even the end of life may reveal this rhythm. Research has found that deaths cluster around the time of the full moon, the period when the numbers of suicides, hospital emergencies and other crises also rise. These are the nights when sirens wail.

Accustomed to our work weeks and financial months, we've come a long way since our ancestors hunted or hid depending on the level of moonlight. Nevertheless—living with twenty-four-hour access to light and in climate-controlled environments—our bodies find ways to remind us of how deeply we remain connected to the cycles of nature.

11

The Years Add Up

B iologists have long known that animals have inborn yearly cycles. Eons of evolution designed them to hunt when the conditions were best, store up fat when temperatures fell, and handle hardships when they were most likely to meet them. When scientists raise animals in a constant temperature, alternating twelve hours of light and twelve hours of darkness, the animals go into hibernation and display annual reproductive behavior as if they lived in the wild. Isolated from the seasons since birth, their bodies never feel the temperature drop nor see the sky darken, but they know what to do and when.

Humans, too, appear to have inborn yearly rhythms. The weather inside our bodies changes to cope with the changing seasons outside. These cycles play out quietly over our lifetimes, altering slightly with each passing year. Year by year, we age and our cycles age and change with us.

The Body's New Year

In society, we celebrate New Year's in the dead of winter, but in biology, our bodies may be more sensible. They celebrate in the spring. In many of our physical aspects, we are restored once winter's winds fall still.

When balmier days begin in spring, many of us feel a sense of euphoria. Our moods turn as bright as the flowers popping up in our gardens. Beginning around mid-March we may find ourselves eating and sleeping less. Our blood pressures and temperatures rise. Around this time, children also experience a growth spurt.

Our immune systems also perk up. By June the number of B-cells, essential for combating infection, doubles from what it was back in January. T-cells, our first line of defense against viruses and tumors, increase in April, perhaps in preparation for the risks we face when we're more active out-of-doors.

Among children, infectious diseases mount a spring assault, with mumps, chicken pox, measles and rubella emptying school classrooms faster than spring fever. Among adults, doctors may be seeing more serious consequences of winter's lower levels of immunities. Breast cancer is most often diagnosed in the spring, with about thirty percent more cases found in May than in December. In men, prostate cancer is likely to be discovered, with a March peak forty percent above the annual average.

Emotional illnesses, too, take their toll in this season. Psychiatric emergencies rise, bringing a springtime surge in hospital admissions for depression. Alcohol abuse and its attendant problems increase from March through June, and suicide shows an April peak. This is the season of joie de vivre for some, but not for all. Chronobiologists suspect that they may know part of the reason.

In spring, and again in autumn when similar peaks occur, some people may not adapt comfortably to the change in day length. Shorter days signal our sleep-wake cycles to accelerate, or in the autumn to slow down, but our temperature cycles may not be able to keep pace. Exceptionally sensitive people may feel out of phase with the world for days on end, as if they are seasonally jet lagged.

These spring troubles fade once summer arrives. We go outdoors more often and get exercise, improving the quality of our sleep. Our metabolism burns more of the calories we eat, rather than turning them into fat, so between May and September we may find it easier to lose weight. Our risk of heart and respiratory problem declines, even in locales where the weather stays fairly constant year-round. We work with our bodies and they work for us, enjoying long summer days. That is, unless the days are too long.

Those who live near the Arctic Circle, where sunlight continues for nearly twenty-four hours in the summer, can be active around the clock. It's not unusual for Eskimos to work and sleep on totally random schedules based only on the weather and hunting conditions. Residents living near the Arctic show lower amplitudes in their daily rhythms, and visitors often experience an eerie sense of timelessness. They may enjoy the freedom to read or work out-of-doors at midnight, but find their sleep disrupted and their waking hours hazy. One manager of an airstrip high in Canada's northern latitudes, for example, set out to do a routine inspection of facilities one "day." He got started only to realize that he'd gone to work at four in the morning.

No matter what the latitude, summer and autumn can also be stressful for those who suffer from allergies and asthma. Winter and spring are kindest to asthma patients, but August begins a difficult phase that will last into November. Meanwhile, the body's immunities rise again, to a peak smaller than that of spring. Then they shift to a winter pattern that provides lower levels of overall protection.

Some of us start sneezing. Others call in sick with the flu. We engage in rituals of hand washing, to avoid the latest bug, and vitamin popping, to combat it. We also resolve to lose weight so that we can enjoy ourselves over the holidays. We've signed on for a losing battle.

Not only have our immunities changed, so has our metabolism. Perhaps to sustain us during harsher conditions, our bodies begin converting more calories into fat. We eat more, and children may experience a growth spurt five times as powerful as the one they went through back in the spring. As for those of us who are as big as we want to be, many join the forty-seven percent who, in one New York City survey, confessed that they gained the most weight during fall and winter.

At least we cannot eat while we're sleeping, and we do sleep more from November through January. We may blame our fatigue on autumn's correction of daylight savings time, which threw us off, or blame it on the lulling syncopation of rain hitting the roof. Actually our bodies' cycles are to blame. Scientists who study this annual phenomenon say it's not exactly like hibernation—since our metabolic functions don't fall so precipitously as animals' do—but humans definitely do beef up and slow down as winter approaches.

This seasonal decline may result in part from how the brain responds to light. As darkness falls each day, our brains begin secreting our natural sleep potion—melatonin. Autumn days are shorter, so melatonin may start flowing earlier. In every species studied so far, the melatonin secretion lasts the shortest time in summer and the longest in winter. Once again we're at the mercy of evolution.

For many people, autumn and winter bring increasing lethargy. One survey showed that half of us feel our energy slump in winter, and nearly a third find themselves less interested in social activities. Yet despite this epidemic of exhaustion, it turns out that winter may be our most emotionally healthy season.

Contrary to pop-psychological folklore, psychiatric emergencies decline during the holidays. We may overeat, elevating our cholesterol level to its December peak, take a drink or two more than usual, and bankrupt our savings in gift-giving frenzies, but we're happier. Once the holidays pass, however, both our spirits and our health take a nosedive.

In January those who are susceptible to migraine headaches are more likely to get them. Those with heart and respiratory problems also face increased risks. Death from all causes peaks between late February and early March, in part reflecting increased numbers of heart attacks and strokes. Our bodies are still fighting all those colds and flu viruses, we're carrying around winter's extra pounds, and our stalwart resolutions to slim down may begin to look like the overly optimistic illusions of New Year's Eve. Fortunately, seasonal help is on the way.

By mid-March, springtime and the body's new year arrive. Our immunities bounce back. We shed pounds. The increased hours of sunlight restore our pep. Crocuses tie bright bows on the gift of spring, and we awaken to discover we're survived another year of both outer and inner weather.

Life Seasons

The year-long *circannuals* are our longest cycles, at least of those discovered so far. For each of us, however, there is a single very long cycle—the longer the better—that marks the passage from infancy to old age. Moving through this gradual pattern, our biological clocks undergo subtle shifts as well.

We notice some of these changes. A woman in her reproductive years may find herself attentive to her hormonal cycles, for example. Later in life, changes in our daily rhythms of activity and rest may be more noticeable, as well as the yearly cycle that makes it harder or easier to lose weight. More than other chronobiological features, our cycles of waking and sleeping shift over our lifetimes. Being aware of these changes can help us age, if not gracefully, at least in better tune with biological time.

Whatever mechanisms guides our inner time, it begins its work even before birth, while we are still in the womb. Seven months after conception, the fetus already shows the rhythms of sleep, and even premature babies exhibit a temperature cycle. These cycles free-run, not yet geared to the outside world, and for a while the baby's rhythms remain out of touch.

Individual clocks may be working, but not yet linked to others. Newborns fed on demand let their mothers know they're hungry about every ninety minutes, but since urinary rhythms take longer to develop, diapers need changing on an entirely random schedule. Newborns sleep sixteen to eighteen hours of every twenty-four, but cycles of sleeping and waking take a while to get organized. A rhythm separating day from night begins to appear around the sixth week, and clearly emerges after the fifteenth, but as many as five months may pass before the infant's rhythms resemble those of its sleep-deprived parents. Naps are popular among the crib crowd, and among exhausted parents as well.

Throughout these early months, the infant's circadian pacemakers gradually begin to link up, both to the outside world and among themselves. Within individual rhythms, the range of highs to lows is more narrow, with sleepiness, hunger or high temperature seldom far off. By ten or eleven months, a baby's temperature clearly follows a twenty-four-hour rhythm, but two years may go by before wakefulness can be counted on during the day and episodes of sleep concentrate at night.

Once the household settles down, the baby's parents may finally hope to "sleep like the babies," but this is impossible. Adults drift for a while before falling asleep, but babies drop directly from wakefulness to full sleep. At six months, the baby still sleeps about fourteen hours a day, and while adults have four stages of sleep, babies start out with only two.

The first stage has no eye movements, and during it breathing is regular. This gradually matures into non-REM sleep. The baby's second stage, the precursor of REM sleep, features rapid eye movements, along with changes in breathing and episodes of sucking. During the first eight months, babies cycle through their sleep stages about once every hour, but a cycle closer to ninety minutes emerges after about a year. The baby is finally ticking more in time with the rest of us.

It can take two or three years for an infant's body to establish fully coordinated daily rhythms, but the essentials arrive first. After about a month its body temperature begins to follow a daily rhythm of highs and lows. Within the first year, growth hormone begins to link up with slow-wave sleep, arriving in pulses as frequent as those in adults but in much larger doses. Babies' bodies seem to know what they need and when, so that infants consume more fat and carbohydrates during spring and summer, then gain the most weight—a survival advantage—in late summer and early autumn. By age two the average child is sleeping about half of each day, an hour or more of this in a nap. Four years later, by age six, most children skip naps to sleep about eleven hours, all at night.

Hyperactive Ten-year-olds and Sleepy Teens

During this time the child's proportion of REM sleep has dropped sharply, and this decline parallels a rise in alertness. As any parent of a ten-year-old knows, by the time children reach this age they are among the most alert creatures on earth.

They're on a normal twenty-four-hour cycle, but they cannot nap, no matter how hard they try. They sleep about ten hours a night, but they're experiencing more frequent and higher bursts of growth hormone than the adults trying to keep up with them. They're growing like weeds, and their inner clocks are embarking on a shift that will leave the rest of us behind. Compared to them, grown-ups are sleep-walkers.

While it seems that young people are going faster than the rest of us, in chronobiological terms their clocks slow down as they approach puberty. For a decade, beginning in the teenage years, the body's measure of a day lengthens to twenty-six or even thirty hours, as if the clock were running slow. If a fifteen-year-old resists going to bed before midnight, that's because his or her body actually thinks it's still dusk. When the rest of the world awakens, the teen remains behind, convinced it's two or three in the morning. They're getting less melatonin, but growth hormone and reproductive hormones are flooding their systems.

Teens and their parents battle through this desynchronization as well as they can, unless the situation turns serious. This can happen when a young person's rhythms free-run, falling so far out of time with the world that life gets turned upside down.

Some teens, at the mercy of a condition called Delayed Sleep Phase Syndrome, truly cannot get to sleep. In DSPS several inner clocks run much too slowly; and as the body lives on its own schedule over succeeding days, sleep comes later and later. One sixteen-year-old, for example, suffered with this problem for two years, managing to synchronize his sleep-wake cycle with the outer world for only one day each month. The rest of the time, he was bounding around at midnight, nodding off in class, and feeling hungry when everyone else was snoring.

Researchers pulled him out of school and put him in the hospital to make sure he got to bed at eleven each night. Gradually his sleep-wake cycle adjusted enough for him go home and hold a part-time job delivering newspapers early in the morning. Over a January vacation, however, he began sleeping late. His body fell out of sync again and could only be synchronized, after months, by strict adherence to an 11:00 P.M. bedtime. By March he was back in school, at least until year-end exams. Then studying until midnight threw his cycles off again, and it became clear that only through the most rigid adherence to a regular bedtime could this young man keep in tune with the rest of the world.

Young people with DSPS may eventually benefit by taking nightly doses of that sleep-inducing hormone, melatonin, to reset their rhythms. For most teens, fortunately, a regular wake-up time can serve as sufficient time-giver to keep the body in tune. They need ten or eleven hours of sleep a night, but when they come up short during the week they can catch up on Saturday or Sunday, putting in more time in REM. Their alertness is lowest in the morning and rises as the day goes on, which may help explain some of the yawning epidemics teachers must tolerate.

By college age, young people sleep an average of eight hours each night, going through regular cycles of REM and non-REM similar to those of older adults. They are also waking up a couple of times a night, but generally they don't know it. They fall back to sleep immediately, an arrangement that older adults may eventually come to envy.

Settling Down With Sun Cycles

It takes until our late twenties or early thirties for our bodies to settle down to a twenty-four-hour day. By this age our temperature and sleep cycles are coordinated with light and darkness, and our energies are on call when we need them.

Unfortunately, this situation does not last. Through a gradual shift during our thirties and forties, we begin spending more of our sleep in REM, where we can be aroused easily, and less in sleep's other states. We may awaken briefly, and notice it, but manage to drift off again. Meanwhile, with more time in REM and less in sleep's deeper stages, we lose the advantage of those major bursts of growth hormone.

It takes longer for the body to repair normal wear and tear, perhaps accounting for the bad reputation middle age has for aches and pains. We find it harder to adapt to schedule changes, and some who have always enjoyed night work may put in for a shift change. We're sleeping six or seven hours at a stretch, but our sleep doesn't seem to restore us as much as it once did.

We feel this way because biological clocks throughout our bodies have begun to get ahead of the twenty-four-hour day. Just as we're looking forward to slowing down after retirement, our bodies begin an acceleration that will shorten our days, disrupt our sleep and blunt our alertness. The clear effects of this emerge only after the age of fifty, but it probably begins earlier.

Studies suggest that a breakdown may occur on two levels, both in how well our inner clocks respond to outside cues and in how well they communicate with one another. If we have difficulty responding to outside signals, such as light, our troubles can be compounded when inner clocks also fail to respond to synchronization signals coming from each other.

One indicator of rhythmic trouble with age has already been found. After the age of eighty, the SCN—the likeliest candidate for a biological master clock—has less volume and significantly fewer cells. Researchers suspect that this decline may be preceded by less conspicuous changes they have yet to discover.

We do know that our reproductive cycles change. Women's monthly rhythms become irregular and eventually cease. Men still have a daily testosterone cycle, but its peaks are lower. The levels of other hormones decline, too, including that of cortisol—the energy hormone—and melatonin. These shifts affect blood pressure and chemical balances, as well as energy and mood. Older bodies also have more difficulty keeping their separate rhythms in tune, with one study finding abnormal relationships between heart rate, temperature and urinary cycle in three of four older men.

Fast Clocks, Slow Bodies

Meanwhile, our temperature cycle undergoes its own transformation. It begins its daily rise about two hours earlier in older people than it does in the young, as if the day had only twenty-two hours. Seniors are up and about earlier, making them better candidates for those crack-of-dawn college class times, but their energies fade sooner too. In chronobiological terms, their temperature rhythm is phase-advanced, meaning it gets started sooner in relation to the cycle of day and night. In addition, an older person's temperature rises only about one to one and a half degrees each day, while a young person's may rise two or three degrees. Since temperature rise prompts an increase in alertness, the range between sleepiness and full alertness narrows.

Naps become more popular. Sleep is precious at any age—we spend about a third of our lives enjoying it—but at no time is it more treasured than in old age. As one recent Gallup poll reported, nearly half of those older than sixty have difficulty getting a good night's rest.

Solid sleep is hard to come by because several features of our inner rhythms combine to prevent it. For one thing, having one's temperature operate on a twenty-two-hour day, when one needs sleep based on a twenty-four-hour one, goes against the grain. It means fighting back yawns through dinner and the early newscast, hitting the sack early only to awaken, as more than ten percent of the elderly do, around 5 A.M. For women, the problem is even more pronounced. Their temperatures begin to decline hours before their husbands' do, and they reach a midsleep low about an hour sooner. Meanwhile, sleep itself has changed.

After age fifty it's common to sleep as few as six hours per night, and many people manage only five or fewer. Rather than our REM periods lengthening as night goes on, they become equally long, and our sleep cycles grow shorter and more numerous. Many old people simply drift in the shallows of sleep's earlier stages, never reaching the deepest stage in a given night. This means that some of us awaken hundreds of times each night, episodes of alertness that may last only seconds or stretch out to a full half hour. This breakdown of the body's time sense might even have health consequences. Studies of insects have found shorter life spans in those whose clocks go awry.

Losing Touch With Outer Time

Our inherent circadian drift can be compounded by social and physical conditions common in later years. When we don't get out in the sun enough or when illness confines us indoors, our melatonin cycle may lose its pronounced amplitude, leaving us sedated in perpetual mental twilight. One small study showed that, on average, the elderly spend about an hour outdoors each day, with women getting even less outdoor light than men.

Women do have more sleep problems in old age, and the extent of discomfort they may face is illustrated by the case of one woman whom researchers identified as having cycles that lost about a half hour each day. If she went to bed at ten o'clock on Sunday night, nine-thirty felt like a natural bedtime on Monday. Following this schedule, the next Saturday night would have her retiring around 7 P.M., risking snapping awake before three in the morning.

Scientists working with her found that, by using bright light for four hours in the evening, they could advance her circadian clock by a remarkable period—six hours. Not only did her cycles of sleeping and waking shift, so did her temperature and her body's secretion of our natural wake-up drug, cortisol. Such studies suggest that getting out-of-doors or using artificial light may help reset the rhythms of the elderly, and perhaps even improve their sleep.

Those who experience physical problems or live on institutional schedules cannot always go out-of-doors, of course, and the illnesses of old age also appear to affect the body's sense of time. Cardiovascular problems and Alzheimer's disease can alter the temperature, melatonin and sleep-wake cycles so completely that day and night become reversed. In some cases the use of light therapy helps correct such desynchronization.

Whether we are healthy or ill, our evolution has fated us to be creatures for whom time itself is limited. The upper boundary of our life span appears to be 115 to 120 years, and some suspect that breakdowns in our cycles of cell division, growth and reproduction move us toward life's end. Our clocks lose accuracy as we approach death, and immediately before we die their rhythms may fade so completely that we drift through time unanchored. Most often life's clock falls silent during the night hours or around dawn on a winter morning. At that moment, time ends for one of us.

12

Is Anyone in Charge Here?

With so many functions depending on internal timepieces—from the nervous system's brief ultradians to whatever slow-sweeping metronome rules the length of our lives—it's natural to wonder how many clocks we actually have. Researchers know of at least one hundred so far, and the full count is not yet in. It appears that we have timekeepers in nearly every organ and type of tissue, as well as inside some individual cells.

Scientists also want to know how these clocks communicate and whether any central mechanisms keeps them in sync. As it turns out, we probably have not one but two so-called master clocks. How well they're synchronized significantly affects how we feel.

In the Business of Time

To visualize our many clocks working together, we might imagine the body as an immense corporation. Different divisions of this successful firm—the BodyCorp—do different jobs. Some place orders and process raw materials—the Hunger and Eating departments—while others transform those materials into products to be marketed by still other divisions—the Thinking and Speech departments. Each division has work areas, and clocks hang on walls everywhere. These timepieces vary in size and shape, from the time clock where folks punch in, to a moon-round schoolroom clock in Accounting, to an elegant grandfather antique—complete with pendulum and chimes—in the office of BodyCorp's president. Despite their differences, all these clocks share one thing: At any given moment, they show pretty much the same time.

Yet some clocks are more important than others. The elegant grandfather on the top floor determines how all the clock-watchers below use their time. If their activities slip off schedule, Grandpa points it out and workers adjust—cutting back hours or working overtime.

In another sense, however, each clock influences the others, including Grandpa. If production problems prevent an order from being ready on time, the folks up in Marketing feel the pinch. Time accelerates for them. Flurries of phone calls go out. Advertising copy is rewritten; television ads are rescheduled. Meanwhile, higher up in the executive offices, time has been slowed. Scheduled projects must be set back. Long-term financial plans become even longer term. As in our bodies, when one clock falls out of tune, the others must compensate.

That's how it goes when everything's working, at least. But if a competitor wanted to sabotage BodyCorp's efficiency, the best plan would be to sneak in at night and set all the departments' clocks to different times. Just as Manufacturing thought it was quitting time and headed for the parking lot, BodyCorp's president would arrive, an event usually occurring at 9:00 A.M. The cleaning crew would start emptying wastebaskets at midday, and folks in the lunchroom would wonder why they were hungry at three in the morning. Chaos would reign.

Such a situation could not last, of course. Workers and messages go from one department to another, so someone would catch on fairly soon. Apparently something similar happens inside our bodies. Chemical or electrical signals pass from one organ to another, alerting different systems that their clocks need to be reset to stay synchronized. In addition, just as the factory whistle blows twice a day—so that not only the factory but the whole neighborhood knows the correct time—our bodies get two time-givers daily. These are dawn and dusk. Although not nearly as predictable as factory whistles, they're considerably prettier and do the job satisfactorily for the real BodyCorp.

Our Daily Time-Check

Just as factories need whistles, our bodies need regular signals from the outside world to keep them operating on schedule. Chronobiologists call these signals *zeitgebers*, a term literally meaning "time givers." The beauty of zeitgebers is that they keep our one hundred-plus clocks from getting completely out of touch with the real world. At BodyCorp, disordered clocks would bring only a loss of profits. In our bodies, when our rhythms lose pace with one another, we may fall ill or, at the least, have trouble sleeping or functioning during the day.

Our cycles can fall out of phase with one another because our internal clocks keep time very flexibly. To see how this works, we can look again at that modern family we met in the first chapter.

The mother's body thought that days were twenty-six hours long. If she free-ran like that in the real world, gaining two hours a day, in less than a week she would be waking when her husband was ready to go to bed and arriving at work at the hour her employer locked up for the night. Fortunately, because she lives in the real day-night world instead of a continuously lit cave, this woman's body resets its clocks every day.

Chronobiologists don't yet know quite how this resetting works, but they have identified a few things—zeitgebers—that make it happen. Light is one zeitgeber. This woman's body probably manages to adjust to a twenty-four-hour world because light enters her eyes each day and tells her brain to turn back that twenty-six hour clock by two hours. It is as if her watch had a light meter attached, so that she did not have to reset it herself. In more scientific terms, chronobiologists say that this woman's biological rhythms are *entrained* by the zeitgeber of light.

Light also entrains the rhythms of animals. Patricia DeCoursey used pulses of light to phase-shift the activities of flying squirrels. In addition, food and activity can act as zeitgebers for animals. If rats find that the rat chow begins arriving twelve hours before it's expected, they will gradually change their inner schedules to accommodate the new feeding time. Social activity, too, appears to provide a zeitgeber for humans and animals. If a creature regularly gets together with its own kind at a particular hour—to squeak, bark or discuss politics—that contact helps reset its inner clocks.

The flexibility built into our inner clocks, allowing them to drift occasionally and be reset by zeitgebers, serves a vital function. If an animal sought food on a rigid twenty-four-hour schedule, no matter what the season, it would be doing so in darkness during half the year. Predators might be out and about; prey might not. If a species were not entrained by light, it would probably not survive.

As for humans, if our bodies were designed to operate on rigid twenty-four-hour days, we would not be able to keep going into the night when survival demanded it, nor awaken before dawn to prepare for hunting. It's also doubtful that a rigidly timed species would have invented night-shift work, jet travel across time zones or even daylight savings time. Flexible, if occasionally inaccurate, inner clocks allow us to adjust to shifts. We can do so because, in evolutionary terms, an inaccurate watch is worth more than Tiffany's top-of-the-line timepiece.

How Every Body Cooperates

While our biological clocks are imperfect compared to manmade timepieces, their ability to entrain themselves to zeitgebers keeps them in time in the outside world. We don't yet know whether every clock can be reset by signals from outside or if only certain master clocks are reset. These, in turn, might reset second-level clocks, and they might reset third-level ones, and so on down the rabbit hole of time.

Tracing the lines of communication among our timepieces calls for some ingenuity on the part of chronobiologists. They can only be certain they've identified a clock if they can isolate it from nearby signals that might keep it on time. They must remove it and keep it alive in a laboratory, allowing it to free-run at its own rate. Numerous experiments reveal that isolated clocks have different ideas of how long a day lasts. Adrenal glands, for instance, follow a daily rhythm slightly shorter than twenty-four hours. Liver cells think a day is about twenty-five hours long.

Isolation studies, like the one Stefania Follini participated in beneath the New Mexico desert, also reveal our separate clocks' schedules. Under constant light and with food available around the clock, animals and humans gradually lose inner synchronization. For a while their internal organs communicate, but the feedback signals seem to fade and rhythms disconnect. After a few weeks below ground, family members' appetite centers might ring dinner bells when their brains were asleep; and sleepiness might arrive thirty, forty-eight or more hours after their last wake-up.

A third way to study separate clocks involves changing their timing. This is possible because genes determine the period of an individual clock. Fruit flies, for example, sing love songs when they're ready to mate, but a mutation on a single gene can produce a specimen out of time and tune with its fellow flies. Using this technique, one researcher deliberately created a fly that put on singing performances every forty seconds instead of the usual one song every sixty seconds. The poor fly probably thought he was oversexed. His love objects surely considered him in too big a hurry.

Genes governing specific rhythms have also been found in mice, chickens, and humans. Thus, members of families tend to display similar rhythms; and some rhythmic disturbances, such as abnormal blood pressure patterns, run in families.

It's All In Our Heads

If individual genes rule our internal cycles, it appears likely that some central pacemaker must keep all our rhythms synchronized. Like that grandfather clock on the top floor at BodyCorp, something must be in charge, or the whole corporation would fly apart. Through a combination of sheer luck and targeted research, one such master clock already has been found.

The first breakthrough came in the 1960s when a researcher noticed that the hypothalamus—a portion of the brain about the size of a cherry—had fibers connecting it to the eyes. Long recognized as playing a role in sleep, the hypothalamus also contained the SCN, a densely packed cluster of only eight thousand cells. The SCN had scientists baffled until they saw that it was also linked, directly or indirectly, to many areas of the body.

If light went from the eye to the hypothalamus, and from there to the SCN, that bundle of cells might act as a sort of transformer. It could relay information about time to the rest of the body. Soon the SCN began to look like a central biological clock.

Experimenters set out to test this theory. They found that mild electrical stimulation of the SCN shifted an animal's free-running period, as if it had seen light at a new time. Destroying the SCN eliminated rhythms of drinking and activity. These cycles did not disappear immediately, however. Instead, they decayed over several months before becoming entirely random. Perhaps, researchers theorized, the SCN merely coordinated signals coming from elsewhere in the body and added the crucial impact of light. Within BodyCorp it might play the role of a negotiator, keeping all departments' schedules in mind while adding information about the world beyond company walls.

Nevertheless, secondary evidence suggests that the SCN has very powerful connections at BodyCorp. Before birth the SCN's parts gradually interconnect, then continue developing to become a fully integrated network. This would correspond to the time it takes for newborns to synchronize to the day-night world. SCN cells taken from older animals respond differently to hormones than those taken from younger ones, which may explain the changes in our rhythms as we age. Blinded animals and some blind people have difficulty keeping in tune with the cycles of light and darkness, which fits with the theory that the eyes send signals to the SCN.

In addition, single SCN cells react to pulses of light and can free-run, following their own daily pattern. Such cells are sensitive to melatonin, too, so they must get information from the body's sleep-wake cycles. As years passed, evidence accumulated to boost the SCN to the status of a master clock. Then, in the late 1980s, two scientists took a crucial step.

Transplanting Clocks

Since genetic differences can cause one member of a species to run slightly faster than another, researchers decided to see what happened if they transplanted an SCN from one genetic type to another. They selected three strains of hamsters. One genetic strain ran on its exercise wheel based on a traditional twenty-four-hour schedule. The second obeyed an accelerated day, one only twenty hours long. The third—a product of crossbreeding between the first two—took to its exercise wheel every twenty-two hours.

Using delicate surgical procedures, the investigators removed the hamsters' SCNs and transplanted them, mixing and matching SCNs between the three. After a suitable time for healing, they timed the animals' activity cycles.

If a twenty-four-hour hamster got an SCN from a twenty-hour brother, it now hopped on its exercise wheel four hours earlier than usual. The reverse also held—twenty-hour hamsters lengthened their idea of a day by four hours. If an originally quicker or slower critter received an in-between clock, it settled down to a fairly sensible twenty-two-hour running schedule.

Switching the SCNs reset the animals' activity in every case. This very small portion of the brain had to be a central clock.

Do We Also Have a Grandmother Clock?

Yet even after those hamsters showed off their new timing, several questions remained. Chief among them was whether the SCN was our only master clock. Lots of evidence pointed to the SCN, but significant clues pointed elsewhere.

Destroying an animal's SCN abolished some of its rhythms but not all of them. Without SCNs, for example, monkeys lose their rhythms of sleeping and waking, but their temperatures still follow a regular cycle. In isolation, humans' and animals' rhythms drift, but not all cycles drift at similar rates.

Among humans, when volunteers are kept under fairly constant light and given wristwatches that show a day that's only twenty hours long, some of the chemicals in their bodies obey the shortened schedule and other chemicals do not. One particularly clever experiment placed volunteers in arctic-summer conditions, with outdoor light around the clock, then switched the timing of day and night. Here, again, some chemicals shifted, but only after four days. Other chemicals required a full eight days to pick up the new schedule.

Converging lines of evidence now suggest that our bodies contain two master clocks. The one already discovered, the SCN, is probably crucial to our cycles of slow-wave sleep. It may also play a role in the secretion of growth hormone and adrenaline, as well as in annual cycles of reproduction. A second clock, whose location remains uncertain, appears to govern our internal temperature, a crucial indicator for alertness and mood. It may also schedule when we manufacture and eliminate certain chemicals, as well as our cycles of REM sleep. Thus, sleep becomes the product of two clocks' input—the SCN for the slow-wave portions and the mysterious "temperature clock" for REM.

The first clock, located in the SCN, is more flexible. When we're exposed to light and darkness on thirty-hour schedules, our SCN-linked functions adapt without apparent difficulty. This flexibility, means that some inner rhythms accept new work schedules or time-zone shifts fairly quickly, taking only a couple of days to get into line.

The yet-to-be-found temperature clock has a harder time making changes, probably because it lacks the light signals the SCN gets. Exposed to thirty-hour cycles of light and darkness, it stubbornly stays on a twenty-five-hour day, refusing to adopt the new cycle. Thus, while we may quickly pick up a new sleep routine, our alertness—which has a lot to do with temperature—lags behind. Chronobiologists call the situation when the two clocks show different times *internal desynchronization*. Most of us know it as feeling off kilter.

Because the temperature clock is less flexible than the SCN, it is more powerful. It prefers to stay put, requiring stronger signals to change its schedule. The ratio of strength between these clocks appears to be two to one, with the temperature clock twice as resistant to change.

Obviously the clocks must work together, but no one yet knows how they do it. If light is a factor for both clocks, one might be set by receiving signals about increasing illumination at dawn; the other may stay on time by perceiving increasing darkness at dusk. Or perhaps a weak chemical link connects the two. It might send messages that collect only slowly, then reach a threshold that can reset one or the other. It's as if BodyCorp's president reset his grandfather clock, the SCN, for daylight saving time. But a pound of triplicate paperwork and a dozen phone calls had to be generated before the vice president reset the grandmother clock, or temperature cycle. Meanwhile, various departments operate slightly off balance, following different ideas of the correct time.

The Maladies of Mis-timing

When the timing inside our bodies is disturbed, we definitely sense it. We may feel depressed, disoriented or even ill. If the two-clock theory is correct, we're experiencing the consequences of their ill-timed relationship.

After resetting our clocks for daylight saving time, for example, we may feel out of sync for a week or so. That's only a one-hour shift. Forcing ourselves to leap over six or even twelve hours can require considerably longer for readjustment.

In the interim before our two master clocks recoordinate, one clock, running on the new time, may get so far ahead that it catches up and "laps" the other. One volunteer, for example, began his stay in isolation with his inner rhythms synchronized on a period of about twenty-five hours. Before long, however, he began sleeping and waking every twenty-nine hours. Nevertheless, his temperature still rose and fell on a daily cycle of about every twenty-four hours.

If his sleep gained five hours on his temperature each day, he would experience a brief period of being in phase once every five or six days. His temperature would be 180 degrees out of phase just as often, halfway between the times the two clocks were in phase. In-phase days would feel great. Those out-of-phase interludes—including most of the time—probably resembled the sensation of sleepwalking underwater.

Since we're not in isolation, most of us stay synchronized by seeing the zeitgebers of light and darkness for at least part of each day. Nonetheless, the modern world allows us to control our light and time environments. We can set an alarm clock for 3:00 A.M., close the curtains at midday, or switch on the lights at midnight. We can swap time zones for the price of an air fare. We can do all these things, but they might not be good ideas.

Wondering about this, one scientist got curious about what might happen if we tried resynchronizing ourselves again and again. Suppose, for example, that a jet-setter zigzagged between New York and Paris every three days, with an occasional zag thrown in for a sojourn in Moscow or Hong Kong.

Reluctant to risk the experiment on human subjects, not to mention pay the price for all those air fares, he set up an experiment with three monkeys and a lot of light bulbs. He arranged for the three to live in cages where the overhead lights went on and off every six hours. They were trained to press a certain lever when a special signal appeared and press another when a bell sounded. Thus, the monkeys spent their "daylight" hours doing work and getting tasty rewards. When the lights were out, they slept.

As soon as the monkeys adapted to a light-dark schedule, however, the researcher changed its timing. If they got accustomed to pressing levers and getting treats from noon to 6:00 P.M., he switched to darkness at midday. As soon as they settled into the evening shift, he made them into night workers or turned on the overheads at the crack of dawn.

Before long the monkeys' work performance deteriorated. After two months one went off the deep end and started pressing levers madly any time the inspiration struck him. A second pressed the levers correctly, but stopped eating the treats. The third was very excitable, which may have cost him his life. He died of a heart attack—a cause of death that is rare among monkeys.

Clock Repairs

No human would willingly live under such conditions, of course. Yet even if we keep our schedules fairly regular, and avoid risky research projects, some of us are highly susceptible to internal de-synchronization, risking fates similar to those monkeys'.

Schedule changes at work may make us feel under the weather for a week or two. A midday nap may throw us off for the rest of the afternoon. If we're particularly sensitive, due to age or genetic make-up, we can feel constantly at the mercy of desynchronization— struggling to concentrate when our bodies don't feel up to it, but wide-awake after midnight. As scientists learn more about which master clock governs which function, and how our clocks communi-cate, they are beginning to tinker with our inner time to solve such problems.

Insomnia affects many older people, for example, because as we age our temperature clocks apparently shorten each day by about thirty minutes. Meanwhile, our sleep clocks hold steady at about twenty-five hours. If sleep is negotiated between the two, the sleep cycle can, day by day, feel nudged forward against its natural timing. Frank, a man in his seventies, may fall asleep around 8 P.M., but then wake up feeling disoriented at four in the morning. His temperature cycles got started early, but his sleep cycle wasn't finished yet.

Among younger people a more common complaint is Delayed Sleep Phase Syndrome, when bedtimes slip later and later. In one study of people with this problem, researchers found that both their clocks ran on days longer than twenty-five hours, but that the temperature cycle ran slightly ahead of the sleep cycle.

This means that Frank's granddaughter, sixteen-year-old Nancy, never heads for bed when her family does. She lives on longer days than other family members do, but her temperature cycle gets ahead of her sleep cycle. With her inner rhythms out of tune most of the time, her alertness drops off around ten in the evening, but she's not tired until three in the morning. When she finally does get to sleep, her temperature has begun to rise, getting ready for her to wake up only a couple hours later. Finally, when her sleep cycle lets her awaken, she is well past her peak alertness for the day.

To correct mild problems with delayed sleep, chronobiologists recommend that young people obey a regular wake-up time. For older people, they urge the opposite approach: a consistent bedtime. For those suffering more serious desynchronization, a few drugs, including melatonin, offer the possibility of resetting our clocks, but a less invasive method uses light and timing.

A New Light on Time

Since light can reset one or more of our clocks, and since we have electric lights aplenty, we should be able to use artificial light to keep ourselves synchronized. The approach is promising, but it's taking a while to work out the details.

Not just any length of exposure to light, at just any old time, will do the job. In fact, the wrong amount at an unsuitable hour might make matters worse. Our inner clocks will only accept certain intervals as proper periods, and these periods can only be provoked if light comes at the correct moment.

To return to BodyCorp, it's as if a national emergency required the company to extend or shorten its workday. A presidential order arrived announcing the new period. Accustomed to eight-hour days, the workers were now told that they would put in only five hours, or perhaps increase their daily responsibility to ten hours. For most workers these would be acceptable periods.

If the announcement required that they work only five hours and seventeen minutes per shift, however, or increase their usual workday to eleven hours, rumblings of a strike might arise. Five- or ten-hour days are fine, but not the other choices. Thanks to union negotiators, BodyCorp's workers might strike a compromise, assigning work for the natural periods nearest to the unacceptable shift lengths. They would settle on either five- or ten-hour days, patterns which come more naturally.

Similarly, our master clocks appear to have periods they're inclined to take, no matter how long external signals say a day lasts. Animals can anticipate meals arriving every twenty-three to twenty-five hours, but they won't start salivating on schedule if the food shows up every eighteen hours or every thirty. Eighteen- and thirty-hour day-lengths are simply not in their master clocks' vocabulary. Similarly, human bodies will accept days that are near the range of twenty-two to twenty-eight hours, but refuse to adapt to other time spans.

If we're exposed to very bright lights, six times the level of normal indoor lighting, we can stretch our days somewhat more. Using such powerful zeitgebers, humans have managed to adapt to days as short as eighteen and a half hours and as long as thirty-one. Under these conditions, however, only the temperature cycle truly adapts. Other inner rhythms break free, going their own ways and adopting their preferred cycle lengths.

The Right Time For Brightness

Aside from staying around the twenty-five-hour range, timing signals must also come at the right moment. It's as if BodyCorp's workers—a cranky lot—were picky about the time of day when presidential orders are permitted to arrive. An order asking for a two-hour increase in working hours will achieve varying impacts depending on when it comes.

If it arrives at midnight, the workers will put in the usual eight hours on the day of the message, delaying adopting the new schedule until another day passes. If the message arrives at 8:00 A.M., however, they'll take on the challenge immediately, calling home to let their families know they'll be putting in an extra two hours that very night. If the message arrives at 10:00 A.M., further from the crucial moment for resetting, they'll stay at work for one extra hour, but not two. The next day they'll accept the full ten-hour day.

Like the unionized workers at BodyCorp, our bodies are fussy about when a resetting message arrives. The timing of light determines whether we'll accept its signal immediately or wait a day.

There appears to be a moment, between 4:00 and 5:00 A.M., that determines what will happen if the lights go on. Light arriving before this will set our clocks backward—deducting hours from our idea of an appropriate day. Light arriving after this switches us forward—adding hours. The closer to this moment the light arrives, the larger the number of hours deducted or added. Thus, the real BodyCorp works more or fewer hours on the day the signal arrives.

Recently one researcher became curious about what might happen if BodyCorp got its light order at the precise moment of tip-over—the instant when the body decides whether to turn its clocks backward or forward. By timing light exposures carefully within a window only eighteen minutes wide, he achieved a startling result: His subjects' clocks stopped. Their bodies no longer showed daily ups and downs of temperature. The level of cortisol in their bloodstreams ceased to vary over the course of the day. For biological purposes, they had stepped outside of time.

Presumably, a jolt of light at the right moment set them ticking again. Meanwhile, the possibilities that could flow from resetting and actually stopping our inner pacemakers suggests intriguing directions for future research.

As chronobiologists learn more about the location of our master clocks and how to alter their settings, they foresee a future when they can deliberately adjust our inner timing. They hope to prevent health problems and help us adapt to the strenuous demands of modern society. Eventually doctors may prescribe light therapy with pinpoint accuracy—writing out chits for "500 watts to be taken at 10:00 A.M. and 3:00 P.M."

Meanwhile, the rest of us must do the best we can. By knowing how inner rhythms affect our work and travel, our family life and health, we can better cope with the desynchronization that is epidemic in our electrified world.

IV

What Timing Means

13

Inner Time Is Money

W hen we meet someone new, one of the first questions we ask is, "And where do you work?" Behind that inquiry lies a deeply defining concept. On our days off we may be parents, stamp collectors, skiers or community organizers, but what we do for a paycheck is inextricably linked with how we see ourselves. In many ways, what we do equals who we are.

In the arena of biological rhythms, work also defines us. Here the crucial question is not so much what we do as when we do it. As creatures originally designed to get food during the day and hide from predators at night, we're now free to work all day, all night or around the clock. If we work in tune with our bodies, we feel well. When we fight their cycles, we pay a price.

The Twenty-Four-Hour Employee

Most of us visualize ourselves as part of a nine-to-five, five-day-week world, but that vision ignores a huge crowd of our coworkers. It overlooks heavy briefcases carried home for the weekend, off-hour phone calls and stopping by the office on Sunday to get the tail end of the week wrapped up. It ignores the transportation industry, where airplanes, trains and trucks operate around the clock seven days a week. It omits our health care system, which cannot schedule illnesses or accidents, and an army of teachers carrying home student papers to work on at night and over weekends. Those we turn to when trouble strikes—fire and police personnel, telephone operators and members of the military—work whenever we need them.

Others serve us on weekends and in the evening—restaurant and custodial workers, entertainers and the attendant at the gas station/minimart. Meanwhile, in both the blue- and white-collar worlds, work hours circle the clock. While the sewing machine operator winds up the night shift, a stockbroker may toil in an office nearby, getting a predawn jump on the London market.

As more of us ignore the cycles of the sun so that we can mesh with the cycles of the economy, the traditional workday and work week blend away into continuous employment.

The Body's Workday

Biological clocks dictate physical and mental highs and lows, so how well we do our jobs depends on what our bodies are ready to handle. To learn about our abilities, chronobiologists have examined everything from the speed and accuracy with which people multiply numbers or deal playing cards to how well they remember random numbers and words. They have measured how capably we recognize patterns flashed on a screen, grip an object or lift a load at various times of day. Such experiments show that early in the day, we're best at thinking and communicating. As hours pass, these capabilities wane, but our physical abilities improve.

Recognizing this seesawing of skills, some theorize that our two clocks control different types of abilities. Perhaps the sleep cycle, which may be linked to the SCN, dominates our thinking abilities. Meanwhile, the temperature cycle, geared to the mystery clock, oversees our physical performance. That's only a theory, of course, and many loose ends need tying up before employers can use it to structure our work flow or schedule our breaks. Nonetheless, as individuals we can allocate the days' hours to capitalize on our strengths and dodge our weaknesses.

If our jobs depend on generating ideas and communicating them, we will probably excel during the morning hours, reaching peak form around noon. These are the hours when attorneys argue their best cases and teachers can best hope to enthrall their students. These are also the hours when a sales pitch is most likely to turn into a done deal; more business contracts are signed over lunch than at any other time of day.

If our jobs demand that we take on a welter of new information, analyze it quickly and make decisions, morning is also prime time. Short-term memory peaks before noon, and we learn most efficiently between 9:00 and 11:00 A.M. We are most alert, and our reading speed skims along at its fastest clip. These are hours to formulate a marketing plan or structure a vacation package, to figure the fastest route through a political maze or design the new office computer network.

Once noontime passes, so does our mental brilliance. Our abilities can vary by as much as twenty percent over the course of the day, and after lunch we're on the downward slope of our intellectual Olympus. Nonetheless, the remainder of the day holds out promise for those who think and talk for a living.

Our reading speed may be slipping, but our ability to comprehend what we've read actually improves. We can remember its message more accurately later because, while short-term memory is a whiz kid in the mornings, our long-term recall improves late in the day. Students would do well to cram for exams right after breakfast, then take them that very evening.

We'd be fools, however, to expect to pass any memory test right after lunch. We'd be lucky to successfully pass anything, including the car in the next lane on the highway. Between 2:00 and 5:00 P.M. we enter one of the day's two "zones of vulnerability," when our ability to concentrate and make decisions diminishes. Accidents rise to their daytime peak. Even if we stay safe at the office, the remaining work hours may look interminable. That's because, based on our inner clocks, time actually has slowed.

Guessing at Time

The first evidence that there might be a physical cause for our sense that afternoons last forever came in the 1930s, and it arrived unexpectedly. Hudson Hoagland, a physiology professor, was caring for his wife, who was ill at home with a fever. He went to the drugstore, and when he returned, his wife accused him of being gone much too long. By his watch, the trip had taken only the usual twenty minutes.

He checked his wife's temperature—104 degrees. Out of curiosity, he asked her to estimate a minute by counting out sixty seconds. An experienced musician, his wife had good instincts for timing, but her estimate of a minute came up startlingly short. No wonder she'd thought he was gone too long, he realized—her fever must be affecting her perception of time. After her temperature fell, he tested her again. Now she did much better at estimating how long a minute lasted.

Repetitions of the experiment confirmed that when her temperature was high, she counted seconds too quickly. When it was low, she counted more slowly. After investigating the phenomenon in volunteers as well, by artificially raising their temperatures, he confirmed that when our temperatures are up, we feel as if time slips by faster. When we're ill with a fever, this pattern gets exaggerated, based on the fact that our perception of time shortens by ten percent for every degree our temperature rises.

During those long afternoons at work, our temperatures are climbing toward their daily peaks. We glance up at the clock, but its hands dawdle. We work a bit more and check again, then double-check against our watches. Either both timepieces have stopped, or we are— we assume—overly eager to punch out and go home.

Our ability to estimate time is most accurate in the morning, especially between eight o'clock and ten o'clock. By midday people tend to think it's earlier than the clock reports. We're fairly good at guessing the hour again around 4:00 P.M., but by day's end we tend to assume it's later than it truly is.

All this assumes, of course, that we're sober straight-arrows while we're at work. Using alcohol, amphetamines or opiates makes time seem to race, while marijuana, LSD and other hallucinogens trick us into thinking that time has slowed to a crawl. Neither approach improves the quality of our work, and they may even deepen the doldrums of the midafternoon slump.

The Late Afternoon Lift

Fortunately, once we escape afternoon's sleepy "zone of vulnerability," some of our mental skills begin to kick back in. These functions appear tied to the temperature clock, improving as temperature rises.

The speed and accuracy with which we add or subtract numbers improves around 6:30 P.M., the same hour at which our temperatures peak. Our vigilance also soars. For example, when Navy recruits were monitored to see when they could best detect and respond to a faint signal amid noise, they turned in their best performances around the time their temperatures topped out.

In addition, afternoon brings improvement in our physical abilities. Morning's ninety-pound-weaklings will not shape-shift into Charles Atlases, but their strength and flexibility are best late in the day. As the afternoon progresses, we require less oxygen to do the same jobs that we huffed and puffed over earlier. If we move furniture or heft heavy boxes for a living, it's comforting to know that between midafternoon and early evening we can grab hold tighter and pull harder. If we work on an assembly line or run machinery, our hands are also steadier and we work at our swiftest clip.

Athletes, too, turn in top scores during these hours. Studies of swimmers, runners and rowing crews show afternoon and evening improvements by as much as thirty percent. Both aerobically and anaerobically, we're geared to deliver our finest as the sun descends toward the horizon; perhaps our ancestors, if they had not yet found food before nightfall, needed to excel as shadows lengthened.

Split Shift Society?

Given the consistency of this morning-afternoon pattern, business managers might want to use our cycles to get the best from us when that best is available. If the Hypothetical Company wanted to move to four-hour work days, it might consider employing thought workers early in the day, then sending in the manual workers for the afternoon shift.

The Hypothetical Company would arrange for executives to come in the mornings. Their presentations and decision-making sessions could be held from 9:00 A.M. to 1:00 P.M., when verbal and thinking skills are sharpest. Around noon the support staff would arrive, since afternoons offer the most efficient hours for repetitive paper-processing tasks and routine follow-up. A few hours later the blue-collar crew reports for work. Taking advantage of the physical benefits of the day's waning hours, they handle tools and machinery, as well as lugging heavy cartons to the loading dock. The Hypothetical Company thus gets the most work for the lowest salaries, and may even use space more efficiently.

A Shift in Time

Of course, most of us work twice as many hours as Hypothetical's employees. Some put in three times as many. In addition, for one in four Americans, the eight-to-five day has been replaced by a nonday shift. Here, too, employers and workers can capitalize on the latest insights about inner time, working with it to improve employees' performance as well as health and safety.

Aside from the day shift, typical schedules include the evening shift, from 3:00 to 11:00 P.M.; and the aptly named "graveyard" shift, from 11:00 P.M. to 7:00 A.M. Since we're daylight creatures, the day shift comes naturally to us. It's also fairly easy for our bodies to adapt to the evening shift; our twenty-five-hour inner clocks seldom mind slightly later bedtimes. Those who sign up for the graveyard shift, however, may feel as if they landed roles in *The Night of the Living Dead*. Our bodies were never designed to live against the sun's cycle.

A worker starting the night shift needs at least a week before his rhythms fall back into line, after resetting themselves by an hour or two each day. That assumes everything goes perfectly. In many cases, three weeks can pass before the change filters all the way through the body's systems.

Extreme morning types, about twenty percent of us, have the most difficulty. The pay may be better after midnight, and the night shift may offer the only slot open, but even high motivation will not make pure larks into night owls. Because their inner clocks are the least flexible, larks' sleep may become disturbed from the very first night. At the outside, they can hope to put in eight months on the new schedule before feeling stressed out, exhausted and near burnout.

Even those who are not extreme larks often find they cannot stick with late shifts. The overwhelming majority of workers requesting a return to the day shift cite sleep difficulties as their main reason. Our natural tendency is to grow more larklike with age, so night work is a young body's game. Many workers in their forties and fifties, some of whom have worked this shift for a decade or more, find that they can no longer tolerate it. Fortunately, by that age they have the seniority to put in for an earlier shift. The transfer may mean a lower salary, but the pay cut is worth it to join the wide-awake world.

The Shift-Work Sleep-Walk

Night work is kindest to extreme owls, since they tend to have longer daily rhythms than anybody else. Their sleep cycles appear more flexible, allowing them to snooze fairly soundly during daylight. The ideal night worker—we'll call him Todd—is a natural owl in his early twenties.

Todd is highly motivated by the paycheck if not by the job itself. We might picture him as just out of college, recently married, and pushing himself to pay off college loans as well as save a down payment for a first home.

Todd can get by without enough sleep, but just get by. If his job is boring, a loss of just two hours' sleep results in poorer performance. As Todd builds up a sleep debt in the course of the week, his motivation declines. By Friday dawn, even if he's about to send off a payment on that loan, life has become a grim struggle. When we're sleep-deprived, we can still do our jobs, but we have difficulty remembering why we should bother.

If Todd switches to hours that are closer to normal on his days off, his body accepts the change almost immediately. Like many of America's seven to nine million night workers, he yo-yos back and forth between partial adjustment at both ends of the spectrum. This means he lives in a permanent state of desynchronization, and various systems in his body have a hard time keeping up.

For one thing, Todd must eat at times when his digestive system is unprepared to handle food. As many as twenty percent of those who begin shift work find themselves suddenly developing sensitive stomachs. Shift workers have two to three times the average risk of ulcers and gastrointestinal problems, and Todd's coworkers also experience more back pain and respiratory problems. Their risk of heart attack is twenty percent higher than that of other workers, and, as a study of Minneapolis police officers found, as many as half of them show abnormal patterns in blood pressure. Todd's female coworkers also have an increased rate of infertility.

These risks come, in part, from higher levels of job stress and emotional pressure. Todd's new wife is likely to be working days, so these two hardly see each other during the week. On weekends the pressure is strong for him to adopt her schedule, or for her to stay up all night on his.

Shift work is hard on household scheduling, and families react to the strain. Men tend to use alcohol to cope with the stress, while women more often use sleeping pills or tranquilizers. These self-medications can further disrupt sleep patterns.

The Search for Sleep

Sleep, of course, is the best antidote to the stresses Todd is experiencing, but sleep does not come easily. Even confirmed night people, following their own schedules, lose sleep when they try to rest in the daytime, typically as much as one night's worth each week. Soundproofing the bedroom may help, but it's still difficult to sleep well, because our urinary rhythms tend to awaken us.

Todd might try compromising by taking naps, but these can do him more harm than good. The body expects uninterrupted spans of rest, and those who nap have greater difficulty sleeping in longer stretches. For Todd, the best pattern would be to sleep in two stretches, each three or four hours long. He may also want to do a bit of remodeling both at work and at home.

The crucial factor for Todd's body is light. When we work under normal lighting, our eyes adjust to let us see adequately, but the brain knows how much light it's getting. Night workers adapt more completely when they keep their work areas very brightly lit, fooling the body into thinking that night is day. The level of illumination needs to be fourteen to twenty-four times as strong as usual indoor levels.

When it's time for sleep, day must be turned into night. That means keeping the bedroom as cool and dark as possible, perhaps blocking the windows with removable foam board, cut to fit and covered with foil. It can also help Todd to eat full, cooked meals for his middle-of-the-night lunchtimes. This may synchronize the body with the zeitgeber of food.

To maintain adjustment, Todd must stay on his upside-down schedule over weekends and, if possible, when he's on vacations. That's a high price to pay, even if the salary rate does improve after sundown.

There is one other option that might appeal to Todd, at least temporarily. Instead of working only nights, he might sign up for rotating shifts. This transfer would allow him to revolve between day, evening and night shifts, putting him on schedule with his body—and his wife—at least one third of the time.

Unfortunately, as recent research reveals, this solution may be harder on Todd than the dilemma it was designed to solve. Working three shifts may not be as rigorous as working three jobs, but, thanks to traditions which are literally backward, it can feel even worse.

Time-Lapse Living

The practice of rotating workers' shifts began because employers wanted to avoid permanently condemning employees like Todd to the graveyard shift. They arranged to rotate crews, regularly moving them to earlier times—first the night shift, then the evening shift, then the day. Weekly changes were most common, with the new hours beginning on a Monday so that workers could use the weekend to get ready. Unfortunately, this backward rotation, pushing sleep to earlier times, is exactly opposite of what the body wants. Even more unfortunately, many industries still use it.

The natural tendency of our twenty-five-hour clocks goes forward, advancing activities to later times. If we want to move backward, we may have to drift all the way around the clock to get into phase again. It's as if two runners—the job and the body—took off at the sound of the starting gun, but went opposite directions. If they'd headed the same way, they'd probably spend a fair amount of the race running near one another and might even run abreast or trade the lead. Going opposite ways, they meet up only twice, halfway around the track and at the finish line.

Counterclockwise rotation requires the body to run this race; but instead of two runners, dozens circle on tracks with a variety of larger or smaller circumferences. The body's many clocks cycle at their own rates, and getting them all to cross the finish line around the same time—a single bedtime, for example—is highly improbable. They stagger in slowly, and they are cranky about it.

After a shift change, our ability to solve problems and make decisions improves for the first couple of days, then deteriorates before improving again. Our skill at simple jobs, like adjusting a screw or catching defects on an assembly line, plummets immediately and takes weeks to recover. This may be because the temperature clock, apparently governing physical ability, is more resistant to change.

Shift changes do the most harm on jobs that require dexterity or manual skill, precisely the ones for which rotations have traditionally been used. Alertness levels of night workers are typically one half those of other workers; and since repeated rotations may lower that level even more, handling tools or machinery can be downright dangerous. Some chronobiologists now wonder if permanent shifts, even night shifts, would prove better for manual workers. To take advantage of the brief improvement in mental skills after a shift change, rapidly rotating schedules might be better in jobs where thinking and memory count most.

That approach sounds good for employers, but employees may see things differently. Working rotating shifts means living at a sleepy crawl. More than half of workers on rotations confess to occasionally nodding off on the job. They also suffer more from stomach problems, and their risk of heart trouble skyrockets, doubling after five years of rotating shift work, tripling after ten.

Like those monkeys that were never allowed to adjust to any schedule of pushing levers for rewards, workers on rotating shifts have their inner time repeatedly scrambled. The consequences can jeopardize not only their health but others' safety as well.

Asleep at the Switch

Sometimes the acts of sleep-deprived workers make headlines. When a presidential commission reviewed the events precipitating the 1986 explosion of the space shuttle *Challenger*, it found that key project managers had less than two hours of sleep the night before. Ground workers exercised poor judgment and made errors, in part due to shift work and sleep loss. Seven years earlier, employees at Three Mile Island had been on the night shift for only a few days, having rotated around the clock every week for five weeks. Crucial decisions were made between 4:00 and 6:00 A.M., during the body's second "zone of vulnerability." In the early morning hours the time required to read meters or answer warning signals increases, and reaction time can fall as much as fifty percent below peak performance. These are the hours when truck drivers have the most accidents, train engineers miss the most warning signals and, it would appear, reactor workers overlook hazards that can risk a meltdown.

We seldom see headlines about the far more common moments when individual lives are held in weary, potentially dangerous hands.

One study of medical interns found that more than a third reported an auto accident or near miss during their internship, triple the number they'd had the year before. Interns routinely work 120-hour weeks, sometimes on thirty-six-hour shifts, and a quarter of them acknowledged having been so tired that they fell asleep while talking on the telephone. Given such exhaustion, and the schedules of nurses on shifts and doctors on call, it is hardly surprising that hospital accidents occur most frequently at night.

Airline pilots, too, take responsibility for lives while living on scrambled schedules. Routes that cross multiple time zones can provoke desynchronization that makes rotating shift work look like a calming routine. Crew members who fly at night lose an average of two hours' sleep each day; and older pilots, whose bodies adjust least well, lose even more.

Older pilots are also the most senior, given first choice of assignments that offer the best vacation allowances. It is a chilling irony that the best perks come with the longest hauls. This puts the workers who are least chronobiologically flexible first in line for desynchronizing schedules, and for the flights demanding the longest spans of alertness.

Whether we are pilots, postal workers or telephone operators, the deterioration from shift adjustment is cumulative. At the low ebb of the day, around four in the morning, someone who is already desynchronized may have been working for more than six hours, having to remain vigilant while doing repetitive tasks. All levers, buttons or switches begin to look alike, and the voice of command at the other end of the phone line may seem to sing a lullaby. We do our best, but we are running a race where evolution rigged the odds.

Sensible Schedules

With what they now know about our natural rhythms, chronobiologists have begun helping industry use inner rhythms to improve efficiency and avoid accidents. Several consulting companies throughout the United States perform analyses of workers' schedules and show how to improve them. They work from a few simple principles.

Most shifts rotate once a week, but three-week rotations appear to be far better. Weekly schedule changes do more harm because workers never quite adjust, or barely reach adjustment before facing the next rotation. A recent experiment with the Philadelphia police department illustrates the advantage of longer rotations.

The department switched to three-week rotations for eleven months, then surveyed the officers to learn if they'd experienced any changes. Most said they felt less fatigued and slept better, and that their families were happier. The officers' use of alcohol and tranquilizers declined by half, and far fewer found themselves falling asleep on the job. This no doubt contributed to the forty percent decline in auto accidents.

These officers also rotated in a clockwise direction, an arrangement that allows adjustment fifty percent faster. Researchers have found that at least one day of rest is required between shift rotations, preparing the body for the new demands it will encounter.

In another experiment, researchers saw that a sensible schedule of rotation may suit workers even better than permanent, nonrotating shifts. Using a sample of about 150 chemical plant employees, chronobiologists put some workers on a weekly rotation and others on a three-week pattern, moving both to later shifts. A third group did not rotate at all.

Significant trends emerged in turnover, productivity, health and job satisfaction. Those on rotating schedules overwhelmingly preferred the arrangement, with the three-week rotation outscoring the weekly one by forty percent. Productivity at the plant went up, and the health of those on three-week rotations improved.

Sunlit Factories

Whatever the shift arrangement, firms that ask their employees to work around the clock can profit by using better lighting. One recent experiment clearly illustrates the advantages.

Experimenters arranged for eight volunteers to report to a laboratory each night around midnight and spend eight hours working under lighting equal to the level of the early morning sun. Once the real sun was well up, around 9:00 A.M., the "workers" went home and slept in thoroughly darkened rooms, awakening at five in the afternoon. Their bodies shifted to their new "day" after only five nights, as evidenced by changes in their temperatures and body chemistry. Peaks of alertness and mental performance shifted, too, indicating that all of their rhythms had been fooled into mistaking night for day.

To put findings like these into practice, one company now sells light fixtures with computer software to control levels of illumination. Since older workers may need more pronounced signals to reset their rhythms, the computerized system takes into account not only the lengths of shifts and the rotation pattern but employees' ages as well.

Even if a company prefers not to provide a sun that rises and sets within its walls, hiring decisions can take workers' natural timing into account. Preemployment testing will identify extreme larks or owls, and the results can influence shift assignments. Medical personnel in company clinics can also be trained to recognize desynchronization and help employees cope with it. In addition, industry and medicine can work together to reduce the risks of exposing workers to health hazards.

Depending on the time of day, various parts of the body respond differently to chemical or biological exposure. One chemical, for example, will kill nearly eighty percent of test animals at a given hour, but only eight percent at another. The body's response to cancer-causing substances, radiation, bacteria and even extreme levels of noise varies dramatically over the course of the day. Since the true risk of such hazards depends on whether the body is geared to cope with them, government and industry may want to keep our cycles in mind when developing safety regulations.

No Time Like the Present

Business and industry are two areas where chronobiological insights have the most immediate applications, and research has already begun to bring change. For those of us who enjoy nine-to-five, five-day weeks, several useful principles have also emerged.

We're all on shifts that change weekly—working forty hours in five days, then enjoying total flextime on weekends. When we let our bedtimes drift later on Friday and Saturday nights, the Monday morning rat race may begin while the racer's body remains sound asleep.

In addition, when we're young we may develop the knack of overriding biological cycles to advance in our careers. As we age and our rhythms become less flexible, our goals and habits may need to become more flexible. As one psychiatrist sees it, pathological arrhythmicity contributes significantly to midlife crises, marital dysfunction and depression.

We're a delicately tuned species, living in a world that has been electrified for only about a century. It just makes sense, when conflicts arise, to remain most loyal to the demands of our first employer—the body that keeps us in business.

14

Other Places,
Other Times

Working rotating shifts has been described as living in a permanent state of "industrial jet lag." For the body's purposes,
moving backward eight hours each week resembles spending a week in
Denver, one in Paris and one in Tokyo. It's an apt comparison,
because "shift lag" has a more glamorous twin—jet lag.

We subject ourselves to this rotation when we set out to get away
from it all, especially from our jobs. Whether we're executives book-
ing flights to distant meetings or vacationers checking bags on the way
to a Club Med, when we buy a ticket to cross time zones, we sign on
for this most modern malady.

Rough Landings

The effects of zipping across time zones have been tested in business
travelers, air crews, actors, chess players, athletes and racehorses.
Even the horses lost ground in the new time zone. It matters little
whether we're outbound or homeward bound, traveling for pleasure or
for profit—despite our best intentions to hit the ground running, many
of us merely hit the ground, hard.

To get some idea of how hard, we can compare jet lag to the simpler adjustment demanded by daylight saving time. In the week after we turn clocks forward or back, automobile accidents increase by as much as ten percent, absenteeism rises at businesses and schools, and many people find their sleep disturbed. That's only a one-hour change. Jet lag—technically called desynchronosis—kicks in when we lose more than two.

One experiment examined the effects of jet lag by using lighting, rather than expensive airline tickets. After a week of monitoring volunteers' normal temperatures, skills, moods and sleep patterns, researchers advanced their schedule of meals and lighting by five to eight hours. A further week of testing revealed the impact: The subjects became irritable and lost motivation. Their physical cycles—including temperature rhythms—desynchronized. A variety of other symptoms fill out the portrait of jet lag.

These can include sleep disturbances, hazy vision and headaches. The jet-lagged may suffer sore throats, upset stomachs, achiness and diminished coordination. Their strength and reaction times also decline, and these are only the physical effects.

Mentally we're equally off stride, with some abilities declining by nearly forty percent. Our brains feel foggy, and we have a hard time concentrating, memorizing or learning. Weary all the time, we feel disoriented and may even experience mild amnesia. If the fine print in tourism brochures mentioned any of this, airlines might have to start paying us to book.

Overdosing on Time Change

Jet lag could be Public Health Enemy Number One except that, like the common cold, it cures itself. Unlike a cold, however, it's not guaranteed to pass in about a week. Recovery depends on the individual, the number of time zones crossed and the direction we went.

About one in twelve of us can honestly claim never to experience jet lag. Of the remaining eleven, five or six suffer it severely. Some people require only a day to recover each lost hour; others may need as many as three days, a full eighteen to adjust to a six-hour change. Those whose daily cycles make the widest swings have the most trouble, with someone whose temperature rises and falls a full two degrees each day facing a bumpier landing than the lucky traveler with a cycle only one degree wide. Jet lag's impact increases with age; and like most desynchronization, it's harder on larks, whose rhythms tend to be more fixed.

The time it takes to resynchronize also depends on the rates at which various clocks drift back into line. The heart rate can take only five days, but even after nine days the temperature cycle may still mistake dusk for dawn. One chronobiologist watched his blood pressure after he'd traveled from Minneapolis to Europe and then to Japan. Two weeks passed before its rhythm resumed a preflight pattern. None of this guarantees that other cycles—digestion, respiration or even cell division—are tuned in to new times yet. Some accelerate to catch up; others slow down and wait for everybody else to circle the track and find them.

Just as our cycles can orbit in different directions, the direction in which we try to move them also makes a difference. Flying north and south causes no problem, as long as we stick to one time zone. Flying west throws our cycles only slightly out of whack, since our natural twenty-five-hour day allows us to add one hour for each sunrise and sunset. Flying eastward is worst. It requires that our bodies do two types of resetting—dropping the twenty-fifth hour and deducting whatever we lost in those missing time zones.

This means that if the whole family gathers for a reunion in Denver, it can take fifty percent longer for the niece from Alaska to adapt than it does for her aunt from Maine. The aunt, taking advantage of the delay on her twenty-five-hour clock, can add nearly an hour and a half per day. The young niece has come eastward. Her mental and physical functions deduct, at most, only an hour each day.

Less Than Friendly Skies

Spending the first days in Denver unable to remember relatives' names or digest potato salad is unpleasant enough. For those who must fly for their living, however, jet lag can present a serious hazard. Pilots, businesspeople and athletes may consider time-zone changes as common as clothing changes, but their bodies get no better at accepting them.

Nearly ninety percent of flight crew members report disturbed sleep, and eastbound flights are known for causing the harshest effects. On average, eastbound crews sleep fewer hours and less deeply than those going west; and after several legs of a multisegment trip, this sleep debt can become cumulative. Among women, frequent flying means less frequent menstrual periods, and the Federal Aviation Administration now restricts female flight crew assignments on transoceanic routes.

The impact of jet lag on those who bounce around the globe doing business is less easy to quantify, since few executives readily admit to falling asleep while trying to draw the bottom line. Athletes' performance is far more public, and the ability to avoid jet lag surely contributes to the home field advantage.

In one study, sixty percent of Olympic athletes acknowledged that the combination of jet lag and the scheduling of events affected their performance. Some Olympic teams hire chronobiologists as consultants to help them capture gold in foreign time zones, and many professional teams incorporate chronobiological principles into rest and practice schedules for games away from home. Following the guidelines used to help shift workers adjust to equally sudden time rotations, they employ preflight scheduling and postflight lighting adjustments to persuade their inner clocks that only the scenery has changed. These methods seem to work. Meanwhile, several popular anti-jet-lag products have no impact except on the wallet.

Taking the Time Cure

If jet lag had been around a hundred years ago, snake oil salesmen would have loved it. Medicine shows might have offered a "Spirit of Landing in St. Louis" and another for landing in St. Paul. Lacking those, but facing a huge potential market, imaginative types have come up with several creative cures, none of which do much.

One can purchase a watch that shows two times, that of the home zone as well as the local time. At least this is handy for scheduling long-distance calls; it does little more. Another watch will accept input for the time at one's destination and the length of the flight, then gradually, during the flight, accelerate or slow down so that both times merge at the moment of landing. It's a nice idea, except that the time showing on one's wrist has no impact upon the time flowing inside one's body.

Of similar appeal, but more widely available in paperback, is a system known as the jet-lag diet. Although some travelers swear by the method—eating only certain foods before, during and after the flight—it has so far resisted scientific confirmation. In controlled studies, volunteers who followed the regime experienced just as many crash landings as those who ate normally. Perhaps the diet's true appeal arises because it partially restricts devotees' consumption of alcohol and airline food.

More effective solutions take the body's relationship to light into account. To fool ourselves, we can partially adjust to the new time zone before departing or go on a strict light diet after we arrive.

One expert suggests preparing for a trip six hours eastward—from New York to Frankfurt, for example—by beginning the adjustment five weeks before takeoff. If the traveler is willing to go to sleep and arise ten minutes earlier every other day, she can get a three-hour jump on her upcoming time shift. She must make up the remainder after landing, of course.

Business travelers may find it inconvenient to reset their bodies before travel, although blaming "jet-lag prevention" for getting to the office three hours late has its appeal. Instead, for important meetings lasting several days, they may choose to arrive one or two days early and put in a few hours sunbathing. It's an investment in alertness, after all. After flying eastward, morning light gives the best return, but westbound travelers benefit most from late afternoon rays. For visits lasting only one or two days, it's best to schedule crucial meetings based on one's peak times at the home office, never switching to the new time.

For those reluctant to pencil in sunbathing time on their Day-Timers, a bit of planning can contribute to smoother landings. After touching down, the most important step is getting outside and seeing local light. In one study, a group of volunteers stayed inside their hotel rooms while others got out and saw the sights and the sunlight. Members of the latter group adjusted much more quickly and probably had more fun. Even if the weather or business demands require staying indoors, making sure the lights are bright during the day and the room darkened at night can bring the all-important temperature cycle into line after only a couple of days.

Chronobiologists are also experimenting with the use of artificial light. One experimenter, facing a walloping fourteen-hour loss on a trip from Tokyo to Boston, decided to plan ahead. By following a structured regime of bright light and darkness for only three days, he delayed his body by eleven hours. Another experimenter found that completely inverting the sun's cycle, by experiencing light at the low point of body temperature, could shift rhythms by twelve hours.

The time at which light exposure occurs, as well as the length of the exposure, is crucial, so it will be a while before we know precise dosages for inner-clock corrections. The harried executive who tries a do-it-yourself approach, staring at the ceiling lamp while shaving, may risk not only cutting himself but conking out at the conference table.

An even faster solution—the sort of quick fix popular in the age of the Concorde—would be to just take a pill. Over the years, physicians have tried various medications to help the body adjust to time shifts. Some seem to work, others cause unpleasant side-effects, and one handled jet lag by giving three neuroscientists severe cases of amnesia. (They woke up in a foreign hotel unable to recall how they got there, cleared customs, exchanged currency or checked in.) No such drawbacks have turned up for the newest pharmaceutical candidate, a substance the body already makes—melatonin.

A Pill for Time

British researchers had the pleasure of checking out melatonin's potential against jet lag the easy way. They got together the funds to take staff members and relatives on a trip from London to San Francisco. The ten men and seven women took either melatonin or placebos before and after the trip.

As payback for the complimentary vacation, the volunteers wore wrist meters monitoring their activity levels. They also had to keep sleep logs, provide urine samples, and record their moods and temperatures every two hours while awake. To stay busy during this fun-filled holiday, every four hours they took tests of their logical reasoning and their speed at crossing out every appearance of a specified letter on a printed page. Spending vacation time crossing out every *e* on this page might not be one's idea of a fun trip, but the project leaders were pleased.

None who took the melatonin displayed significant jet lag. All but two of those given placebos felt a lot foggier than San Francisco's climate should have made them. Melatonin apparently cured jet lag in one hundred percent of those taking it, probably because it shifted their temperature cycles to the new time.

The full consequences of taking this wonder drug have yet to be explored, but if such results hold up, airport gift shops may someday offer bottles of "Mela-perk" and "Lag Away." Melatonin has already been used to regulate the daily rhythms of blind people, whose inner clocks cannot be set by light. Eventually we may combine doses of this hormone, telling the body that it's dark outside, with exposure to pulses of light, prompting the biological day to begin. By judiciously timed use of these two tools, we might escape the time boundaries our inner clocks set for us.

There are some theorists, however, who are less than sanguine about the potential uses to which melatonin might be put. If this substance cures jet lag, should it also be used to cure shift lag? Could those on rotating shifts be tempted, pressured or even required to take melatonin so that they can perform more efficiently on the job? Melatonin dosing is a step somewhat short of genetically engineering a better employee, but some wonder if it's a step in the same direction.

As for jet lag, it's been blamed for everything from failed international negotiations and disappointing athletic performances to family feuds. Yet until melatonin, or some other magic potion, proves its promise, we can only try to plan ahead. Before and after travel, we can sleep and wake by the numbers and enjoy getting out to soak up the sun.

15

A Time to Embrace

In love—and lust as well—Ecclesiastes had it right: There truly is a time to embrace and a time to refrain from embracing. Many of us notice our sexuality changing over our lifetimes, but few observe subtler, equally significant changes affecting desire over the course of the day, the month and even the year. While love is far more than biochemical clockwork, the act of love takes its cues from our sex hormones. They, in turn, rise and fall in response to numerous internal clocks.

Love's Lifetime Clock

The walloping hormonal surges of adolescence put an end to childhood's innocence, and to many parents' illusions about how long innocence would last. Boys will be boys and girls will be girls, and they do so when their bodies say it's time. For boys, in the second decade of life, testosterone begins rising in the bloodstream, particularly at night. Girls begin menstruating between the ages of ten and sixteen, and this evidence of fertility usually arrives in late autumn or early winter. As their cycles continue, menstruation most often begins between 4:00 A.M. and noon, and women whose periods start during these hours tend to menstruate for fewer days than those who begin in the afternoon.

As they mature, testosterone fuels both men's and women's sexual fires, but a man's body produces ten to fifteen times more of this substance than a woman's does. In men, testosterone displays a clear daily cycle, dropping to its low in the evening and early night, then beginning to rise after midnight. By eight or nine in the morning, testosterone peaks, an average of twenty percent higher than it was the night before. Women may also have a daily cycle of testosterone secretion, but the lower levels in their bloodstreams make such a rhythm difficult to detect, at least so far.

As a man ages, his pattern changes. Younger men experience higher morning peaks, their average levels more than twice those of men over the age of sixty. When researchers took repeated blood samples from seventy men, aged twenty-six to ninety, living on identical schedules and eating at fixed hours, they found that this decrease was unrelated to lifestyle or routine. They could also assume it bore no relation to the availability of female company—these seventy monks lived in a monastery.

Cycles of Desire

Neither men nor women display any inherent weekly cycle in their testosterone levels, but men's bodies do reveal one of two rhythms lasting several days. Some show higher testosterone levels every three to five days; others experience testosterone peaks every twelve to eighteen. Women's testosterone is geared to their menstrual cycles, and if they're married, their husbands' bodies tend to mirror these monthly rhythms. Among married men, testosterone rises around the time of the wife's ovulation.

This makes biological sense, since nearly half of all women feel most interested in sex at midmonth. During the three days surrounding ovulation, a woman is more than twice as likely to experience orgasm, perhaps because her sense of touch is at its most responsive. She can hear and taste with finer discrimination than at other times, and her sense of smell is especially acute.

From such data we might assume that the likeliest time for lovemaking would be on the fifteenth day after the woman's menstrual cycle, around 9:00 A.M. But we'd be wrong.

Despite our inner clocks, social forces are often more powerful than biological ones. An interesting finding about our sexual schedules illustrates this. While our bodies may expect us to make love at midmorning and midmonth, desire also depends on luck, locked doors and whether the boss expects us at the office. Among those who work outside the home, lovemaking most often occurs during testosterone's daily low—around eleven at night. As for the day of the week, unless most women ovulate on Saturdays or Sundays, we must be putting our bosses' demands before those of our bodies. We most often have sex on weekends, with Sunday morning being an especially lively time. In such matters it appears that convenience overrides biological demands.

So we'd be wise not to call our friends when we assume they're reading the Sunday funnies. It would be especially indiscreet to call on the weekend when the newspaper carries announcements of Thanksgiving Day sales. This is so because our desires also follow an annual schedule.

Seasons of Love

Full summer moons may seem romantic, but so are rain on the roof and gusts rattling the windows. Although we buy more contraceptives in the summer—perhaps hoping for summer romances—we actually need them more often after temperatures begin to fall. Testosterone follows an annual schedule, with the highest levels occurring from September through November.

When medical students were asked about their love lives, they reported their highest rates of sexual activity in October—perhaps the happiest hunting time for one who wants to marry a doctor. We generally fall into testosterone troughs between January and April. These students skipped sex and caught up on their sleep most often in February.

Confirmation of this active-autumn, lacklust-winter pattern comes from a less appealing statistic—the infection rate for sexually transmitted diseases. Gonorrhea, which takes two or three days to show its symptoms, is most often diagnosed in August. It appears least often in March. The symptoms of syphilis take nine to ninety days to reveal themselves, and are reported most frequently between August and December, less frequently between January and June. AIDS, the newest and most devastating sexually transmitted disease, has yet to reveal an annual cycle.

When we consider evolutionary imperatives, this annual cycle of sex makes sense. It served our species' survival to cuddle up inside caves when winter winds began to blow, then to emerge to enjoy the gentle zephyrs of spring. Like animals, we may also have used yearly rhythms to help our offspring survive.

One researcher found that some women are fertile only during the late autumn and the spring. As for men, the sperm that they ejaculate at these two times of year are better swimmers. During winter a greater proportion of a man's sperm will be abnormal, and during hot summer months its concentration in his seminal fluid may decline.

If we want to avoid pregnancy, taking contraceptive precautions surely makes sense during the autumn months, when those medical students, at least, are having more sex. We would also be wise to prevent conception during springtime. Despite testosterone's autumn peak, we actually conceive children most often right around the time of the vernal equinox—around March 21. A second peak in conception occurs about six months later, when the sun once again crosses the equator. At both these times the earth experiences approximately twelve hours of sunlight and twelve of darkness each day.

Romantic Times and Temperatures

When researchers went back to the nineteenth century and checked birth patterns, they began to suspect that this perfect balance of light and darkness may be part of what signals our hormones' cycles. Looking at records from 166 regions worldwide, they subtracted nine months from the recorded dates of births. North of the equator, conceptions were most frequent during the spring equinox, with a second peak in autumn. South of the equator, the higher peak appeared six months out of phase, in autumn, while the lower one came in spring. For those living well north or south of the equator, the difference between the minimum and maximum was as much as threefold.

At first it looked like our reproductive cycles were based on the length of the day, but a closer look revealed that temperature also plays a role. Before 1930, day length seemed to exert the strongest influence. Yet around that year the pattern started to change. Thanks to Thomas Edison and the Industrial Revolution, more people began working indoors. Birth statistics since 1930 show that we are more likely to conceive when the temperature stays between fifty and seventy degrees. These temperatures are more likely, of course, around the equinoxes.

Thus, if we want to conceive, we might want to check the weather report. Assuming all the signs are right—our estrogen and testosterone levels up, ovulation on schedule, the hours of light and darkness equally divided, and both the day and our desires feeling balmy—we might succeed. Yet even when conditions are ideal, some couples wish to have a child and find that they cannot. For infertile couples—numbering as many as 2.5 million in the United States alone—chronobiology is beginning to suggest solutions.

The Fertile Moon-Month

Many factors contribute to infertility, and part of the solution now appears linked to our physical rhythms. The scheduling of sexual activity, availability of light, and awareness of timing may make a difference in whether or not a woman becomes pregnant.

Women whose menstrual cycles are 29.5 days long—the precise period of the moon's waxing and waning—are far more likely to be fertile. In fact, a woman with such a cycle has a ninety-eight percent chance of conceiving if a healthy sperm arrives at the right time. A woman with a longer or shorter cycle may have difficulty getting pregnant.

When one researcher asked women how often they engaged in sexual activity, she found that those having sex with men at least once a week were more likely to experience 29.5 day cycles. Having sex twice a week or more often further increased the odds for this classically fertile cycle. Celibacy or intermittent sex activity correlated with shorter or longer cycles, and probably less fertile ones. It now looks as though regular sexual activity may act as a time giver for women's menstrual cycles.

Light also appears to play a role. Curious about the similarity of the menstrual cycle to the moon's cycle, one scientist decided to find out whether there was any relationship. Women with classic 29.5 day cycles, she discovered, were two or three times more likely to start menstruating during the light half of the moon's month. They most often began two or three days before the moon was full. Far fewer began menstruating when the moon was dark. Since ovulation generally occurs halfway through the menstrual month, the new moon probably correlated with the time when the majority of these 29.5-day women ovulated.

Additional evidence for the role of light in women's rhythms comes from an entirely different approach. Some women have unusually long cycles, menstruating as infrequently as every fifty-four days. Long cycles may provide fewer windows of fertility, but when researchers left a one-hundred-watt light burning by their subject's beds for five nights during midcycle, the length of their cycles gradually became normal. This technique could eventually help couples who want to conceive.

Yet even when natural methods fail, our inner rhythms may provide a boost to conception. Of seventy-nine attempts at *in vitro* fertilization, the first four pregnancies came with eggs released between 10:00 P.M. and midnight. Perhaps, if one wants to get pregnant, that 11:00 P.M. cuddle is a good idea.

No one's saying yet whether we should make love once a week or twice a week, with the lights on or off. Chronobiologists do believe that sleeping with the light on, during the right time of month, may eventually offer a simple, noninvasive method of birth control. In addition, since our biological cycles have so much impact on our ability to conceive, it is hardly surprising that they also orchestrate much of what happens after conception occurs.

When Two Bodies Share Time

Once pregnancy has begun, a new set of living clocks begins discovering how to keep time. We typically think of this as happening over a period of nine months; but since our bodies read moon cycles far better than they do date books, these nine months are not, strictly speaking, nine calendar months. The average pregnancy lasts 266 days—nine moon-months.

Over this span, the mother's daily rhythms change. While she sleeps, the electrical activity of her brain and the pattern of her eye movements alter, with eye movements becoming most frequent around the eighth month. The daily rhythm of her blood pressure may shift, reaching higher peaks in late afternoons. Meanwhile, the rhythms of her heartbeat signal her fetus, setting up patterns reflected in a new heartbeat.

Animal studies suggest that, even in the womb, the basic rhythms of our offspring echo those of their mothers. When scientists change the daily schedules of pregnant rats, for example, the inner clocks of their unborn pups soon change too. If the mother's master clock, the SCN, is destroyed, the pup's fetal rhythms are also disrupted.

Among both animals and humans, nutrients and hormones pass through the placenta at set times, providing the fetus with time givers from outside. Inside that doubling and redoubling ball of life, no cell is an island. Numerous processes signal what happens and when, and a human being takes form based on a series of clocks and maps.

At prearranged instants, undifferentiated cells find distinct identities—becoming brain cells or liver cells, cells that will shape a hand and others that will eventually see the world as eyes. The heart takes form, and its rate increases steadily as months pass. Blood pressure declines until around the fifth month, then begins a rise that will continue until delivery. Finally, after nine moon-months, a new life enters time as we know it. Even this event has its appointed hour.

Birthing Hours

Back when our mothers had to evade predators long enough to give birth, it made sense that nighttime was the right time for labor. A very dark night, like that of the new moon, provided advantage, but a full moon might also help a creature see to survive. Nowadays hospital staffers prefer working day shifts, no matter what phase the moon is in, but any doctor who chooses obstetrics would do well to be a night owl. She might also want to check the moon's cycles before buying season tickets to the symphony.

Apparently we still prefer to be born during new or full moons. As for the time of year, babies conceived near the vernal equinox begin arriving in December. The child of autumn's equinox, and testosterone's peak, most often picks a birthday in the last week of July.

Labor itself usually starts during the sleepiest hours of everybody's biological day—between 1:00 and 3:00 A.M. Actual birth occurs not long afterward, typically between three and four in the morning. Fortunately, fewer of us choose to enter a sunlit world. If a labor is going to be long and difficult, or if it will end in a cesarean section or a stillbirth, it will most likely have begun during daylight hours.

This pattern of birthing at night may not suit mothers, anxious fathers, or cab drivers on the night shift, but only doctors appear able to do anything about it. When labor is induced, birth usually occurs during prime hospital time—10:00 A.M. to 5:00 P.M. Whether this timing owes more to basic convenience or biology, no one's saying.

As for the rest of us, nature chooses its own sweet times. Evolution acted like a democracy, with only those born at the most inconvenient hours living long enough to cast genetic votes. This clockwork cunning let generations of mothers survive to later tell us how we kept them up all night being born.

16

We Are When We Eat

I f some clever inventor wants to cash in on dieters' dollars, he might advertise a special kitchen clock, perhaps in *Weight Watchers Magazine*. Instead of being marked with numbers, the clock face would show pictures of foods.

Grains and starches would predominate during the morning. High noon might feature a seafood salad or a hearty soup. The afternoon and evening times would whet the clock-watcher's appetite with pictures of protein snacks—a wheel of cheese or a chicken breast—but the clock, being solar powered, would shut down when night fell. A sleepy eyelid might fall over its face, since darkness is the time for fasting.

This menu-clock could be as helpful as a calorie chart because, whether we're chowing down on everything in sight or fasting to shed pounds, our bodies follow a fixed pattern of hunger and satiety. Our digestion gears up at preferred times, expecting specific foods based on a clockwork routine. Even those who, due to illness, no longer have stomachs, feel hunger on a predictable schedule.

Getting Credit for Eating

Thanks to the efforts of hundreds of college students, we now know a great deal about how inner timing governs hunger and metabolism. As any teenager's parents know, young adults are renowned for their grazing habits. They'll eat most anything, at any time, and they'll leave the empty container in the closet or the refrigerator for mother-magic to replace. To support their voracious appetites, college students will accept paltry wages, show up promptly if only they are promised housing and food, and endure procedures the thought of which would make older subjects go pale.

Isolation experiments are especially popular on campuses around exam time, when a student likes nothing better than to hole up in a climate that's controlled, with a free and uncontrolled supply of food, as well as books, lecture notes and a working Walkman. Little matter that the lights stay on twenty-four hours a day or that experimenters are monitoring everything from the exact number of potato chips Joe College eats to how often he uses the bathroom. The pay is right and the academic situation desperate.

When they are put in isolation like this, Joe and his sister Jane drift away from sleeping a portion of every twenty-four hours. Instead, they nod off every thirty, thirty-six, or even forty-eight hours. Yet even though they stay awake longer, Joe and Jane eat the same number of meals. Accustomed to three meals a day, they will still eat three times, meals spaced further apart and in some cases with more generous portions.

Although they may eat bigger meals, however, isolation subjects typically lose weight, because they're expending more energy between meals. It's a radical method of dieting, but it may have appeal as a last resort.

In addition, neither the timing nor the size of a meal dictates when Jane or her brother will become hungry again. Chemicals flowing in their brains tell them that. These chemicals also determine what they want to eat, reminding their bodies to meet the need for expendable energy during more active times and to store up strength when food is scarce. Thus, a calorie means something different to the body depending on the time of day, the day of the month, or the season of the year.

Joe College, a science major, sees a calorie as a fixed quantity— "The amount of energy required to raise one kilogram of water a single degree centigrade," he will answer on his upcoming exam. His body takes a different approach. Biologically, a calorie is a unit of energy that's bigger or smaller depending on the position of the sun. The same macaroon that his body might blow off in activity at ten in the morning looks like a last valuable resource twelve hours later, when the need for sleep necessitates hours of fasting. Similarly, because nature's grocery shelves are overstocked in spring and summer, an eclair eaten in April offers easy-come-easy-go calories. In September, when the body expects the coming winter to make food scarce, the same eclair must be hoarded on the hip or pot belly. We'd do the same thing if supermarkets traditionally stayed closed from October to March.

The Hunger Cycle of the Sun

As they cram for those exams or do a few laps around the lab to stay in shape, the rate at which Joe's and Jane's bodies use energy follows a set cycle. Whether they've eaten or not, their metabolism peaks in early afternoon, burning the most fuel. As one landmark study demonstrated, the calories we eat before this are more likely to be spent as energy. Those eaten later may be turned into fat.

Seven people agreed to eat only one meal each day, either in the morning or the evening, for one week per mealtime. The meal provided two thousand calories, no more and no less, and during the first week they ate it as breakfast. This turned out to be the fasting week, since all seven lost weight, an average of slightly more than one pound. During the second week they ate their two thousand calories as dinner. This feasting week caused all but one to gain weight, about a pound apiece.

On average, Joe and Jane do consume about two thousand calories per day, in a range that may dip as low as fifteen hundred for Jane and rise to twenty-seven hundred for Joe. They eat twelve to fourteen percent of these calories as protein, and three to four times as much in carbohydrates. The times at which they crave carbohydrates—for example, cereals and sugary snacks—versus those when they're hungry for protein—a hamburger or a slice of cheese—vary with the hours. This happens because their brains expect different energy demands at different times.

In the morning, even before we awaken, our brains send out two messengers: cortisol—that energy potion—and a hormone called norepinephrine. If these two walked into a restaurant, they'd order every sweet roll, cereal, and fruit plate on the menu; they are crazy for carbohydrates. Show them a cheeseburger and they'll turn green. Carbohydrates provide the instant energy our bodies think we need before the sun gets very high overhead, during our intensely active mornings. Our metabolism puts them to maximal use around noon.

By that time, the brain has added a new messenger to the team— serotonin. Serotonin has a more balanced view, and it orders a more balanced lunch. It welcomes that protein-rich hamburger and also appreciates the fat contained in the chef's secret hamburger sauce. Suspecting that before shadows lengthen we'll need protein to rebuild muscles, as well as fats to fill fat cells, serotonin damps down our desire for carbohydrates. It likes having a good time, but also puts something away for tomorrow.

Thoughts of tomorrow are paramount for our evening hungers, and to provide for it the brain sends a new messenger—galanin. This is the chubby sort of customer the dessert chef loves to see come in the door. Galanin adores fats, loves them so much it will suppress cortisol, which might have opted for carbohydrates, and sit down to the table with protein-loving serotonin. They will feast on a well-marbled steak, finish the sour cream but leave the baked potato, then call for pastries, pudding and an ice cream sundae.

Despite its bad nutritional habits, galanin is actually trying to help. The sun is headed for the horizon, and nature's restaurant should be closing for the night. We need to store up energy to make it through the hours of sleep and to keep ourselves going in case food is scarce in the morning. Galanin means well, but in our world of twenty-four-hour restaurants, it may have outlasted its usefulness.

Galanin's potential to do harm can be aggravated by two factors: Our sense of taste is sharpest in the evening, and we're free to eat and enjoy because we're unlikely to be otherwise occupied at work. As evening falls, we have both motive and opportunity to get around to eating, with the consequence that some of us get very round indeed.

If Joe and his sister wished to use their time in isolation to lose more weight, they might request that lights be turned out for a portion of each day. This would throw off the experiment, of course, but it would certainly help pounds melt away.

The rest of us could take a radical approach to maintaining a reasonable weight by moving to locales without electricity. There, it would be too dark to find the refrigerator at night, assuming we had refrigeration. A more reasonable approach requires balancing our intake of calories against our output of energy. This means that we, like those college kids, can give in to morning's carbohydrate cravings, but must resist evening's feeding frenzy for fats.

If we postpone eating in the morning by skipping breakfast, by the time we get around to a meal we're playing carbohydrate catch-up. Later in the day we'll get more calories than our bodies will be spending, as well as adding fats we're unlikely to use. Like those research subjects who ate only dinners, we're putting away for tomorrow, and again for tomorrow and tomorrow.

Skipping and Shifting Meals

The three little piggies of norepinephrine, serotonin and galanin—assisted by the hungry wolf of cortisol—turn up at set hours based on where our brains think the sun should be. This schedule overrides any plans we might make for ourselves.

If Joe passed all his exams, for example, graduated and got a job as an astronomer—on the night shift, of course—he would still be affected by when his body thought he should work and eat. While some digestive functions readily accept a new mealtime in the middle of the night, others buy in only after all the body's cycles have made the switch.

For animals, a shift in mealtimes has pronounced effects. Food acts as a zeitgeber, providing a signal that causes other functions to change swiftly. The jury is still out on how completely a change only of mealtimes alters human clocks. To learn whether food was a zeitgeber for us, one experiment looked at the daily cycles of hospitalized patients.

Unable to eat normally because they were in vegetative states, the patients were nourished by intestinal infusion. Researchers shifted the timing of feedings and monitored whether other cycles shifted. After four weeks they found that providing nutrients to the body only at night inverted the temperature cycle, making temperature rise at night when it should be falling, then decline during the day. For most human functions, however, the impact of changing mealtimes is less significant.

When Joe no longer eats at noon but gets to the lunch table at midnight, his digestive tract may be trying to get a few hours' rest. His brain may be secreting chemicals to tell him to sleep. Like workers on rotating shifts, jet-lagged travelers and even harried, hurried types who often skip meals, Joe becomes more prone to stomach trouble. Since he's sleeping when he should be eating, his stomach shows up for work and finds itself unemployed.

This disruption is a continuing problem for those who work night shifts, and Joe's new eating habits reflect what happens. Particularly during the work week, he eats fewer calories than workers on the day or evening shift. Even chowing down on weekends fails to make up the difference. Instead, throughout the week, Joe and his coworkers drift into a pattern of "nibbling," snacking at short intervals around the clock. Most of these snacks are carbohydrates, as if desynchronization tells the brain that wake-up foods are continually necessary. Even during his day-sleep, Joe may awaken and reach for a donut or a bowl of cereal before snuggling in again to get some rest before the sun sets.

Summer Savories, Winter Woes

Our daily cycle of hunger and metabolism has counterparts in the varying desires our bodies express for food during the month and the year. The monthly cycle is particularly pronounced for women, and the yearly rhythm can seriously disrupt the eating patterns of those who are most sensitive to light.

Jane, for example, may notice that her desire for certain food rises and falls over the course of her menstrual month. Treats that can be snubbed one week look irresistible the next. One study monitored the eating habits of three hundred nurses and found that food cravings, especially a hunger for sweets, struck most often in the days right before menstruation began. Women who suffer from bulimia, causing them to overeat and self-induce vomiting, most frequently binge in the five days before their periods begin.

Those who have eating disorders are also subject to an annual cycle. Many anorexics find themselves most depressed and unwilling to eat in the spring; while bulimics may binge, feeling out of control, in the autumn and winter. These unhealthy patterns represent a cycle that all of us experience to a lesser degree.

Jane and her brother take in the most calories in August, as days begin to shorten. Their bodies also want more of these calories in the form of carbohydrates. Autumn and winter are the times when meat and potatoes, especially those potatoes, look delicious. Home for

Thanksgiving, Joe and Jane find that the turkey appeals most for its stuffing and for the sandwiches they can enjoy the next day.

Unlike children, whose bodies bulk up in spring and summer, adults do so in fall and winter. One recent study surveyed nine hundred people to learn whether they experienced a seasonal pattern of weight gain. Seventy percent did; and among them, nine out of ten gained in the winter—an average of twelve pounds.

Meanwhile, during shorter days, these young people's bodies are better at turning food into fat. Joe keeps the gain in check by working nights, but Jane is beginning to stretch a few seams. Her summer and winter wardrobes can be a full two sizes apart.

Autumn's anticipatory weight gain parallels what happens with animals. In the laboratory, animals can be made to gain or lose weight by manipulations of how long the lights stay on or off. Perhaps Jane could equip her home with a rheostat and a timer, tricking her inner cycles into believing that spring and summer—the times when we shed pounds most easily—go on forever.

Yet even if we cannot live in an endless summer, and even if menu-clocks are not yet available, at least knowing about inner cycles can help us maintain a healthy weight. Nature designed us to be solar-powered creatures, and those bedtime snacks consumed under artificial lights contradict inborn schedules. Like laboratory animals, we are fattening ourselves based on cycles of light.

The day may come when we can pop a serotonin pill to help us decline a high-carbohydrate dessert, or boost our cortisol in the evening so we feel like jogging it off. But until slimmer living through chemistry arrives, we must settle for using the most potent neurotransmitter known to science—our will power.

17

When It's Time to See the Doctor

E ven if we eat our meals by the menu-clock, work and sleep by the sun's cycles, and exercise to stay both in tune and in time, occasionally our bodies do run into trouble. We pick up a virus and start sneezing. We come down with a flu and must take a few days off. Or we live with one of many chronic conditions—annoying allergies or serious asthma, diabetes or susceptibility to headaches or digestive problems. In addition, as we age, we become more vulnerable to such health risks as heart disease or cancer.

Inner timing plays a role in all of these conditions. Some illnesses show up predominately at certain times of the day, month or year; others arrive at any time, but display rhythmic patterns in their symptoms. Our bodies' cycles can play a role in what goes wrong and also dictate the best times to treat it. Whether we resign ourselves to ride out an illness, take medications or undergo surgery, inner clocks determine how our bodies respond.

Downtime

It's ironic that we describe ourselves as "catching" a cold or the flu.

136

In truth, though we run, we cannot hide, and these bugs catch us. Apparently they're better at chasing us down when our immune systems lower their guards in autumn and winter. At least this is the case in the northern hemisphere. North of the equator, illness and death from infectious diseases peak between December and March; south of it, they strike most frequently between June and September.

If a cold or flu catches us, the runny nose or congestion is usually worst in the morning, but it's afternoon and evening when we feel most dragged down. Our energies begin their drop toward sleep, and the weariness brought on by illness piggybacks on the sleepiness caused by melatonin. It's as if the infection hits the brakes and melatonin puts the body in park.

What time we start running a fever depends on what caught us. Fevers brought on by bacterial infections most often begin in the morning, between 5:00 A.M. and noon; those caused by viruses send our temperatures soaring in the late afternoon and evening. In addition, the medicines we take to combat infection can be more or less effective depending on the time of day. Some antibiotics, for example, reach their highest concentrations in the blood around noon, even though we may be taking them all day. As for when we'll recover, a flu takes its own sweet time to pass, but we can count on a cold lasting about a week, no matter what we do.

More Lasting Troubles

Not all health problems confine themselves to a single episode, of course. For many people, chronic conditions make health a relative state—tip-top when the problem is not acting up, only so-so when it's being troublesome. Such conditions include allergies, asthma, ulcers and the tendency to get severe headaches. While these problems are chronic, their frequency and severity are far from random. As the many parts of the body work together in time, our inner schedules aggravate symptoms at some hours, banish them at others.

Most physicians perform diagnostic tests during mornings, for example, but this may be the worst time to test for skin allergies. During clinics' prime time, our skin is on its best behavior. When patients are given scratch tests at various hours, they develop the most redness and rash around eleven at night, at a time when allergy clinics are locked up tight. In fact, the skin's responsiveness can be one hundred times more pronounced at midnight than it is earlier in the day.

Thus, if we're allergic to the wool blanket snuggled against our neck, we may itch all night. If we're allergic to the dust in that blanket, however, we're more likely to wake up sneezing before the alarm goes off. Respiratory allergies act up in the morning, perhaps because this is the time when our bronchial tubes are least fit.

Hay fever blocks our sinuses and gives us the itchiest noses just when we're getting set for the day. Then it strikes again in the evening when we're winding down. During midafternoon's lull, some sufferers can mow their lawns or enjoy walks under pollen-laden trees without a sniffle. If we take antihistamines to quiet the symptoms, our body's cycles also determine how long their effects will last. An allergy medication taken at 7:00 A.M. will stay in the body for as many as seventeen hours, but the same pill taken at 7:00 P.M. may have left the system within nine.

Allergies are annoying, but a more serious respiratory condition—asthma—is potentially lethal. As many as ten percent of us suffer from it when we're children, and the problem affects about half as many adults. In some cases an allergic reaction triggers asthma, but in others physical or psychological causes bring on the classic symptoms of wheezing, coughing and breathlessness.

Asthma is often worst at night and in the morning, with most attacks coming between 2:00 and 7:00 A.M. Asthma trouble is seventy times more likely at 4:00 A.M. than it is twelve hours earlier, and one study of nearly eight thousand asthmatics found that a third of them were awakened by symptoms at least once a night. More than half woke up at least once a week.

During the night, everyone's breathing ability declines by twenty to forty percent, but asthma amplifies this decline to fifty percent or more. This exaggerated cycle may set off the bronchial spasm of asthma. When those with asthma are isolated in a dust-free environment, then exposed to dust at various points in the day, their responses chart this cycle. During midafternoon and early evening, they hardly react. At eight in the morning and eleven at night, the same dose of dust can cause wheezing and coughing that lasts for hours.

Aware of these rhythms, some physicians now give their asthma patients meters with which they monitor their breathing over the course of the day. The resulting profile can reveal how well therapies are working. The best results seem to come when asthma medication is taken in the morning, with a supplemental dose, for some, in the afternoon.

Women asthma patients can benefit by knowing that an attack is much more likely just before or during menstruation. All asthma sufferers face increased risk in late summer and early fall, with attacks eighty percent more frequent in August than during the low point of asthma's annual cycle, in March.

Using Inner Clocks to Control Flare-ups

It's difficult enough to manage a schedule or plan a trip when we're feeling well, but those who deal with frequent flare-ups of ulcers or arthritis can find themselves at the mercy of their bodies' whims. For them, chronobiology is attempting to offer a few guidelines about timing activities to avoid pain and reduce discomfort.

The burning pain of an ulcer strikes when stomach acids come in contact with vulnerable tissues lining the digestive tract. These acids, powerful enough to break down food, damage delicate tissue. In most people the intestines secrete a mucous coating, allowing destructive acid to safely flow past, but for ulcer patients problems may arise in both the amount and the timing of this cycle.

Our bodies schedule acidic secretions for the times when they expect us to have eaten, providing more at night—to deal with the presleep meal—and in the morning—to break down the first meal of the day. Those who have ulcers secrete more gastric acid overall, and their intestines may also fail to follow the right schedule for building up a protective coating. Thus, pain comes most often and most intensely at night and in the early morning.

Medications can be keyed to the body's cycle of anticipating food. Taken once a day, in the evening, they can provide protection when the ulcer is most likely to act up, while also reducing side effects.

Diabetes also results from problems in how the body handles food, but in this instance the difficulty lies in the pancreas. By producing insulin, the pancreas enables cells to use glucose, the component of food that serves as our primary energy source. If we get too little insulin, we also get too little glucose, and this can cause weakness, fatigue and, in severe cases, unconsciousness and death.

Like the body's need for other substances, our need for insulin varies over time, with the greatest demand coming the late morning and early afternoon. We need far less at night. For most people the pancreas makes sure there's enough insulin in the bloodstream, but diabetics can have very low levels, or none at all.

By monitoring their glucose levels around the clock, diabetics may find the best rhythm for maintaining their insulin level, helping them control their disease. In addition, because our need for this substance varies over the course of the year, the body needs less in some seasons. Diabetics experience fewer problems during the summer, but their reactions tend to be most severe in the winter, particularly between December and April.

Another chronic illness that displays a daily cycle is arthritis. In fact, arthritis's cycle is so distinctive that the timing of its symptoms—pain and stiffness in the joints—indicates its underlying cause. In rheumatoid arthritis, symptoms increase when our immunities are at their peak, usually around seven or eight in the morning. This occurs because the body's immune system attacks soft tissues in the joints. Osteoarthritis, on the other hand, results from years of movement wearing away the lining of the joint. It tends to feel worst during the afternoon and evening.

Based on these cycles, researchers are experimenting with the timing of medicines to lessen arthritis's painful symptoms, as well as reduce side effects. One study found that by scheduling drugs correctly, patients could tolerate four times the usual dose and double their medications' effectiveness.

Mistiming in Our Brains

When our stomachs ache or our joints feel on fire, we can still push past the pain and force ourselves to function. When our heads ache, however, or if the cells inside our brains lose their rhythm, the consequences affect not only our activities but our ability to act at all. Epilepsy and susceptibility to headaches do just this, making those who have them unable to function until the episode passes.

Day and night, our neurons fire in controlled cycles, keeping us in contact with the surrounding world or permitting us to sleep peacefully. Those with epilepsy, however, experience periods when their brains fail to regulate these rhythms. The cycles run chaotically out of control, causing a one- or two-minute electrical storm that overtakes consciousness.

For most epileptics, seizures are linked to a particular time of day. Forty percent are diurnal, or daytime, epileptics, while another twenty-five percent face the highest risk at night. Daylight seizures typically come soon after the time of awakening, as the brain shifts from its sleeping pattern to one of alertness. Individuals' patterns are so consistent that even those who suffer no seizure on a particular day still show abnormal brain rhythms at the hour when they are typically at risk.

Awareness of these cycles can help patients and doctors carefully time antiepileptic medication. In addition, epileptics may want to avoid the desynchronizing rigors of work on rotating shifts, since fatigue increases the chance of a seizure.

Fortunately, epilepsy is fairly rare and can usually be controlled with carefully timed regimes of medication. Far more common is the discomfort of headache, an ailment so prevalent that it causes an estimated two-thirds of us to miss work at least once a month. The most severe headaches—migraines and cluster headaches—occur when the rhythms that affect the brain's blood supply fall out of synchronization.

Migraine headaches, which affect one in ten people, occur when the nervous system fails to properly control the size of the blood vessels serving the brain. At first these vessels narrow, cutting down the brain's supply of oxygen. Then suddenly they expand, irritating nearby nerves. Migraines most often begin in the morning, when our blood pressure is beginning to rise for the day. Those who have them awaken with pounding, throbbing pain that may last for a few hours or even for days.

Women most often experience migraines around the time of menstruation, during the same phase of the month when they are at greater risk for epileptic seizures. In addition, migraines occur most frequently on weekends, but they tend not to strike during the first part of the week, particularly on Mondays.

Even more painful than migraines are cluster headaches. These bring severe pain on one side of the head, especially around the eye. Cluster headaches typically begin at night, about one or two hours after sleep begins, during the time when the first REM cycle usually occurs. A single cluster headache can last a half hour to forty-five minutes, then retreat, only to return again and again. The cycle may continue nightly for weeks, months or even years.

The hormonal rhythms that govern our sleep and alertness may be involved in cluster headaches. Patients who get them have unusually low levels of melatonin at night, and their cortisol, which brings daytime alertness, begins to flow sooner. Recently researchers have begun exposing cluster headache sufferers to bright lights, seeking to reset these mistimed hormonal rhythms.

What Works, and When

Light treatments may eventually cure cluster headaches, but for most chronic conditions medication provides the only relief. The timing of that medication can be crucial, because for some drugs, the same dose that is safe at one hour may be lethal at another. Even when a drug is given around the clock, intravenously or as a time-release capsule, its effect varies because the body's systems frequently change what they're doing.

Suppose, for example, that an athlete is having muscle spasms and needs a muscle relaxant. Unfortunately, the best drug causes significant side effects. Too much of it may relax his muscles too completely and lower his athletic ability. If he takes too little, he'll cramp up when he needs to be in top form. In addition, if this muscle relaxant concentrates in his kidneys, it can do damage.

A specialist in the new field of "chronopharmacology" can help plan this athlete's treatment. Aware of the times of the day, month and year when muscles are most relaxed or subject to spasms, a chrono-pharmacologist can fine-tune medication to the body's rhythms, increasing this drug's benefits and dodging side effects. At certain hours muscles react differently to particular chemicals, so it may require three pills to uncramp them at 8:00 A.M. but only one to do the job at noon. In addition, since the rate at which the kidneys process their contents varies widely throughout the day, it's important to allow this drug the shortest possible stay there. A dose taken at dinnertime might reach the kidneys at bedtime, then remain there, causing harm, until eight in the morning. A dose taken with breakfast might be out of the body before noon.

Meanwhile, the athlete wants his medication to work hardest while he's in the game. What he needs is a pill he can take once a day and trust to modify its effect depending on several clocks—not only the game clock but those governing systems inside his body. Chronopharmacologists are developing tools to do precisely this.

Time-release pills for asthma and ulcers can be taken once a day and deliver most of their medication overnight, when these conditions are at their worst. For other conditions implantable pumps provide an option. These devices contain programs that change how much medicine is delivered depending on what's happening elsewhere in the body. The approach makes sure that, despite the ebb and flow of inner cycles, enough of the drug stays in the bloodstream. Side effects can be avoided by playing into the body's rhythms for eliminating potentially harmful chemicals. The potential for kidney damage from one antibiotic, for example, varies depending on the season of the year, so dosages should take this rhythm into account.

For medications that do not require a constant level in the bloodstream, the body's natural timing can also offer a boost. Sleeping pills taken at night, for instance, piggyback on the sleep-inducing effects of melatonin. Taken in the morning, when the body's cycles rise toward alertness, these pills have little effect unless they're taken in large, potentially dangerous doses. Even vaccinations, given to prevent illness from beginning in the first place, may be more or less effective depending on the time of day when we get them.

The Payoffs From Timing

Gradually, as insights emerge about how our clocks fall out of rhythm and how to get them in sync again, chronobiological approaches are taking their place along medicine's long-held beliefs about the body's stability over time, its homeostasis. Infectious diseases can be treated in coordination with their natural time cycles. Chronic conditions can be kept in check by scheduling medications for best times of the day, month or year. Yet some problems—like heart disease and cancer—cannot be lived with. Nor will they pass on their own. For life-threatening conditions such as these, knowledge about inner time is becoming essential to treatment.

18

When Inner Clocks Fall Out of Time

A living body, like a business or a family, only works if all parts have a similar idea of what time it is. In the body, it is especially important that everyone agree on how quickly time passes. If an organ loses touch with time, the others that depend on it face severe stress. When even a single cell accelerates its clock and duplicates too often, it can destroy the body that gave it life.

In both heart disease and cancer, inner cycles lose their rhythm. Researchers hope that by learning how mistiming affects our hearts and our cells, we will catch problems sooner, treat them more successfully and combine timing with technology to discover cures.

Matters of the Heart

We call love's failure heartbreak, but when our hearts actually do break, they more often break stride—slipping out of sync with other cycles in our bodies. These cycles affect our blood pressure, the blood itself and the entire cardiovascular system. Since each feature obeys its own schedule, our heart rhythms must mesh with other cycles to stay in tune.

144

Perhaps first among these is our blood pressure. It provides a basic barometer to reveal the weather inside our bodies. Even a mild elevation in blood pressure can double our risk of death before retirement age, and a moderate elevation, perhaps by twenty-five percent, can triple that risk. The old homeostatic standard, based on a spot check, puts normal blood pressure in the vicinity of 120/80 for an adult. But because each factor affecting this reading—the force of the heart's contraction, the size of arteries and the amount of blood moving through the body—has its own rhythm, what's normal varies widely depending on the time of day or the season of the year.

A fifteen percent rise from morning to evening might be no problem; a ten percent rise from one day to the next might spell trouble. Some people have hypertension only at night, when their doctors aren't checking; while others suffer what is called white-coat hypertension, their pressure shooting up the minute someone in a white coat enters the examining room. Since some people may register high numbers on more than one office visit, but show acceptable ranges if they're checked frequently throughout the day, relying on spot checks to diagnose high blood pressure can be about as accurate as checking one blade of grass to see if the lawn needs mowing.

A more thorough approach uses newly developed portable blood pressure monitors. With one of these, a patient can be checked every fifteen minutes, day and night, for as long as forty-eight hours. A built-in recorder logs the readings, charts the pattern of change and may even point to a cause. If the hypertension is caused by a kidney problem, for example, its daily pattern is less pronounced than if the problem originated in the cardiovascular system itself. From the different amplitudes and times for highs and lows, doctors can refine a diagnosis.

Since high blood pressure does its damage over a lifetime, we want to detect potential danger as early as possible, and timing patterns can now help predict hypertension before it begins. Newborns from families with a history of high blood pressure show distinct daily and weekly patterns, and in older children the range and timing of the heart rhythms identify those who are likely to develop this problem in years to come. Alerted by such early warnings, individuals might want to change what they eat or how much they exercise to prevent, or least slow down, the damage.

In addition to elevations in blood pressure, episodes of chest pain can signal damage affecting our hearts. This pain is angina, and it happens when not enough blood reaches the heart muscle. How much blood the heart gets depends on several factors, each with its own cycle, and so angina follows a daily schedule too.

About thirty minutes before we get up in the morning, our blood pressure rises. The blood itself also clots more easily, yet our hearts' arteries have less flexibility to accommodate these changes. This can give the heart nearly thirteen percent less blood, and angina is more likely to strike early in the day. Thus, some heart patients find exercises or treadmill tests impossible to do in the morning, but later in the afternoon, when the heart works at peak capacity, they manage just fine.

Those who use nitroglycerine to control angina reach for their pills most often early in the day, and drugs to prevent the blood from clotting also have their largest impact during these hours. As for exercise, those who want to shovel snow from the walk on that December morning might want to keep angina's daily rhythm in mind, and its annual cycle as well. These chest pains tend to worsen in the winter, even in places like Texas, where the weather stays mild.

Angina's winter peak parallels the pattern for the greater difficulty it foretells—a heart attack. These life-threatening emergencies occur when blood fails to reach part of the heart muscle for so long that cells there die. This may lead to loss of the muscle's rhythm, heart failure and sudden death. While heart attacks occur year-round, their frequency peaks during the winter months, striking most often in January and February no matter what the climate. They are also more likely on certain days of the week and at certain hours.

Thursdays and Saturdays are dangerous days for heart attacks, but Monday is worst, with a rate forty percent higher than on other days. Even for those with unvarying schedules—living on the regular regime of retirement homes—heart attacks show a similar pattern. As for their daily rhythm, a heart attack is twice as likely between eight and ten in the morning as it is in the late afternoon or evening.

While scientists don't yet know all the factors shaping these patterns, emerging insights point to likely suspects. A combination of cycles, including morning peaks for blood pressure and clotting, probably contributes to daily rhythm. The reason for the weekly cycle remains a mystery, but the pronounced annual peak may come from seasonal changes in the concentration of cholesterol and hormones in the blood.

The cycle taking place in individual cells may also play a role. Each heart cell knows its rhythm, a twenty-four-hour cycle that persists even when separate cells are kept alive in laboratories. This rhythm varies with each individual, so after a heart transplant, cells in the donor's heart may temporarily remain as much as two hours out of time with surrounding tissue.

As we learn more about how rhythms affect our cardiovascular risk, new insights can help lower it. High blood pressure can be treated when it's high—not necessarily around the clock. Drugs to prevent blood clots can be fifty percent less effective at certain times of day, so homeostatic approaches that keep a constant level in the bloodstream may not work. For those whose hearts are failing, research has begun on a pacemaker designed to adapt to rhythmic changes. Such a device could smooth out an irregular heartbeat by using rhythmic scheduling along with feedback from timing cues coming from elsewhere in the body.

Even those who have no cardiovascular problems can benefit from knowing their hearts' best and worst hours. No matter what we're doing, our hearts deliver more power per pulse beat if we avoid overexertion and stress during riskier times. We can schedule exercise, heavy lifting or that request for a raise later in the day, when hearts may not be softer but at least they're stronger. Our hearts are the inner clocks that we can hear and feel most easily. They're also the ones whose alarm bells we least want to hear ring.

Cells That Run Too Fast

The threat of a heart attack is frightening, but for many people a diagnosis of cancer is even more terrifying. As the largest segment of the population—the baby boomers—enters the cancer-risk years, our evening newscasts bring reports of statistics showing its rise, word of celebrities who are fighting it or have lost their lives, concerns over toxins or foods which may cause it, and news about the latest search for treatment or a cure. On the most basic level, cancer is essentially a problem of timing, so those who study our inner timing are deeply involved in this search. They are finding that the body's rhythms are intimately involved in cancer's beginnings and may also contribute to its lethal spread.

In organs where cancers begin—the lung, stomach or breast, for instance—cells replace themselves in carefully timed cycles. None of us is the same person he or she was a decade years ago; on the cellular level, we are collections of ten million-million replacement parts. This system works well as long as all the human body's ten million-million cells keep dividing on schedule. Yet if even a single one becomes confused about how often to duplicate, a cancer begins. The offspring of that cell also lose track of time and divide too frequently. Their descendants do the same.

Given the odds of one cell misreading a biological clock, it's not surprising that our bodies create a few tiny cancers each day. Following their own daily schedule our immune systems apparently catch almost all of these. For one in four of us, however, the immune system will let a tiny cell cluster get by. Its cells will continue dividing and will also have the capacity to travel elsewhere in the body, settling in there and dividing again. And again and again. Eventually the cancer grows large enough to interfere with a vital function, causing an organ to fail, ultimately, causing death.

Since, at present, medicine does not know how to remove a cancer once it has slipped past local defenses, we need a better understanding of the timing of cells' divisions. This may make it possible to catch aberrant cell cycles sooner, or perhaps prevent their divisions from speeding up in the first place. Researchers are also looking for ways to boost the power of the treatments we have by working with the body's inner time.

Our Smallest Clocks

Cells normally divide in predictable patterns tied to the time of day, week, month or year. Cancer cells, since they started out just like all others, remain tied to the rhythm natural to their locale, but they interpret time differently. To them, one day may feel like two or even three days, so they duplicate every twelve or even every eight hours rather than every twenty-four.

In cancer's early stages most errant cells stay fairly close to their normal neighbors' schedules. Recent research suggests, however, that as the tumor grows, an increasing number of the cells in it stray from healthy timing. Thus, if a tumor held only one hundred cells, with fifty of them thinking a day was twelve hours long and another fifty believing the right day length was only six hours, the short-day group would divide twice as often. They would soon outnumber their more normally timed neighbors. If one of this group's offspring cut the day's length in half again—to three hours—it would gain yet another, tragically temporary, survival advantage.

Chronobiologists suspect that something like this happens inside the world of a tumor. Early-stage, less malignant cells retain a time pattern closer to the tissue in which they arose, but later-stage tumors fall further and further out of touch with normal time. Some theorize that fast-growing, aggressive cancers may have lost all cellular rhythmicity.

Occasionally a signal from outside the body, if it comes at the wrong time, can cause that first cell to lose its regularity. Studies of animals find that radiation, such as that which comes from X rays, can be more harmful during the day than during late night hours. The harmful effects of substances that cause cancer—carcinogens such as tobacco smoke or asbestos—may do no damage at one hour but be quite dangerous at another.

In addition, patterns in the rise and fall of our immunities can make the difference in whether that original cell's offspring take hold. Immune cells appear to be most prevalent in the bloodstream around the time we awaken in the morning, and their number also varies by season. Our antitumor immunities reach their peak in spring and early autumn, but they are lower in summer. Winter marks their low point in the year.

This annual cycle may be mirrored in seasonal patterns for diagnoses. Of two types of testicular cancer, for example, one appears most frequently in the summer; another reveals a winter peak. Hodgkin's disease, when it arises in young men, most often occurs in the spring. Prostate cancer is most often diagnosed in the late winter and early spring, with a March peak forty percent above the yearly average. December is the month when the fewest cases of breast cancer are diagnosed, but that number rises by thirty percent in May. The waxing or waning of our immune systems may play a role in the start and growth of these cancers.

The Twelve Percent Epidemic

Only a few years have passed since researchers began looking into whether the body's cycles affected cancer. Already many separate clues are coming together to suggest a coherent picture that may help improve diagnosis and treatment. To see how much difference timing can make, we can consider one patient—Noreen. Noreen's life is changed in the spring of her forty-eighth year when a routine mammogram and follow-up tests identify a small tumor in her breast.

Perhaps one of the first things Noreen would be told is that she is not alone. She has become that one in every eight women who will be diagnosed with this disease in her lifetime, a twelve percent rate so alarming that one city has classified this cancer as a public health epidemic. Noreen may also quickly come to understand that her general good health and lack of risk factors had no impact on whether or not she would face this crisis. Seventy percent of those diagnosed with breast cancer had no risk factors, and so far her tumor appears to be at an early enough stage that it has not affected her health.

Noreen would find herself alone, however, in the solitude of having to make difficult decisions that could affect her survival. To make them well, she will want to know as much as possible about her tumor and how it is related to her inner timing. Researchers understand a great deal, but the results of their work are not yet available to most patients.

Self-exams and mammography remain the most reliable method for diagnosing this cancer, but new tools are on the way to provide even earlier detection. Because breasts have daily, weekly and monthly rhythms in their surface temperature, the abnormality in Noreen's breast might have been discovered earlier by monitoring this cycle. A breast affected by cancer shifts from a twenty-four-hour temperature rhythm to a shorter one, perhaps twenty hours. Meanwhile, the breast beside it retains normal temperature patterns. Early work on this principle led researchers to develop a surprising device called the chronobra, which the patient could wear to monitor these differences. Recently a related tool has arrived on the market. A woman can use the Breast Thermal Activity Indicator at home, placing two fiber pads inside her bra, to detect the heat differences cancer causes. Some expect that this device will be to breast cancer what the home pregnancy detection kit has become for pregnancy.

A breast that contains cancer also shows abnormalities in its monthly temperature cycle and women with breast cancer, as well as those at high risk, have abnormal daily and annual rhythms of certain hormones. These indicators may give early warning of risk and may even reveal the presence of a tumor. In addition, healthy women at high risk for breast cancer appear to have a distinct rhythm in that hormone that times our sleep—melatonin.

The normal nighttime peak of melatonin, usually highest around 2 A.M., is as much as fifty percent lower in breast cancer patients. This pattern appears to grow more pronounced as the size of the tumor increases; although whether this is a cause of the cancer or an effect, no one yet knows. Some studies have found that melatonin can slow down the growth of tumors; perhaps it encourages the body's immunities, or perhaps it acts directly on the cancer. Eventually this naturally cycling hormone may be used to fight cancer, not only breast cancer but also melanoma, where its daily rhythm also appears disrupted.

Someday we may be able to monitor the breast temperature cycles of those at high risk for this cancer, as well as their hormonal rhythms. These readings would help determine the best treatment for a particular patient.

As for Noreen, she must make her decisions based on the limited information she can obtain. So far, her tumor is small and appears not to have spread. She must decide whether to have only the tumor removed, in a lumpectomy, or undergo more extensive surgery, a mastectomy. After surgery she will also have to work with her physicians to decide whether to have radiation, chemotherapy or both.

Whatever Noreen decides, her physicians will want to incorporate awareness of her body's timing into her treatment. For one thing, the time of day at which she has surgery can dictate how much anesthesia she needs. Studies of animals have found that the largest doses are needed when the animal is normally most active, but a large dose risks a potentially lethal overdose. For humans, most surgeries are performed when we would ordinarily be active, during the day; no doubt because these are also the surgeon's most alert and capable hours. That's an important advantage, yet anesthesiologists need to keep in mind that biological rhythms can be as significant for safety as the patient's age, sex or weight.

Annual rhythms can also affect anesthesia's potency, and the surgery itself can shift the daily rhythms of hormones and other physical factors. After her operation Noreen's body will require several days to get back on schedule.

Noreen chooses to have a lumpectomy, removing only the tumor, and her surgery is scheduled for a morning a few days later. It will occur one week before Noreen's menstrual period is due. This aspect of timing, too, may play a role in her long-term survival.

Numerous studies have looked into the question of whether the time of month when a woman has surgery changes the likelihood that her breast cancer will be cured. The results are stirring controversy in medical circles around the world. One study of forty-four women, for example, found that those who had surgery just after ovulation, past the middle of their monthly cycles, were four times more likely to live cancer-free for ten years or longer. Surgeries at other times of month brought a quadrupled risk of having the cancer return later. Other studies have divided up the month differently, looking for the ideal time. Answers are still coming in, but for the most part, they agree that the weeks just after ovulation, but before menstruation, are safest.

Why this is so may hinge on two factors. When a tumor is being removed, it might shed cells into the bloodstream, tiny cancers that may lodge elsewhere and begin growing. Since breast tissue is dependent on estrogen, and many cancers arising in the breast remain estrogen-dependent, a high level of estrogen in the blood stream may help such cells survive. A woman's blood contains the most estrogen during the weeks directly following her menstrual period, the time when breast cancer surgery appears most risky.

In addition, a woman's immune system changes throughout the menstrual month. Perhaps the cells that are effective against breast cancer are stronger or more prevalent in the weeks after ovulation. Exactly what is happening is not yet clear, but many surgeons are now checking their breast cancer patients' menstrual cycles before scheduling their surgeries.

Using Time as a Shield

Noreen has timed her surgery well, exactly between ovulation and menstruation. After it is over, she also receives encouraging news. The cancer was indeed small, and no evidence of its spread was found. Now she will follow up with radiation of the nearby area, in case some cells traveled locally through her lymph system. For an extra margin of safety, she also chooses a brief course of chemotherapy.

Since the effects of radiation and chemotherapy depend on whether or not cancer cells are dividing, timing also matters for Noreen's further treatment. Her doctors will want to target her tumor and the nearby tissues to take advantage of the pattern of cell divisions. Here, too, temperature cycles can help.

Thanks to cancer cells' mistiming, the temperature cycle of a tumor can be markedly different from that of the tissues around it. One experiment, for example, compared the effect of treating tumors when their temperatures were high with the results for tumors treated at other times. In a matter of weeks, the cancers that were irradiated at their temperature peaks shrunk by seventy percent. Tumors treated at other hours showed only a thirty percent decrease. The immediate effects were impressive, but the long-term effects proved even better. Nearly ninety percent of the patients treated at the nonpeak temperature had recurrences. Only forty percent of those treated at the temperature peak recurred—less than half as many. In other studies, giving radiation at the tumor's highest temperature typically doubled patients' two-year survival.

If Noreen's doctors know the temperature cycle of her tumor, timing her radiation treatments correctly may mean that any remaining cancer is destroyed as quickly and completely as possible. Careful timing can also benefit Noreen when she finishes radiation and begins her chemotherapy.

Chemotherapy works by destroying cells at certain stages in their life cycle. To do this, doctors wish they had a drug that worked like a delicate scalpel to eliminate only cancer cells. Unfortunately, the drugs now available work more like broadswords. Instead of snipping away only cancer cells, they destroy any cell that's dividing. Cells involved in our immunities, digestion, skin and hair growth divide often, just as cancer cells do. Healthy cells, too, are killed, and that brings both short- and long-term side effects. These include nausea, vomiting and hair loss, as well as dangerous reductions in the body's immune defenses.

If timing could be used as a shield, to protect the healthy cells but kill those with cancer, these side effects would be far less unpleasant and dangerous. So-called best times have already been found for at least twenty anticancer drugs, hours during which the body's rhythms protect normal cells—because they're not dividing—but provide no such protection for a cancer.

To treat ovarian cancer, for instance, patients often receive two drugs, both extremely dangerous. The best results come from giving one in the morning and the other in the evening. On this schedule, patients' immune systems stay stronger, they experience less nausea, vomiting and diarrhea, and they can take larger doses more frequently. If the timing is reversed, however, giving the morning drug at night and vice versa, a patient's immune systems can be harmed so severely that she must delay or even stop chemotherapy.

For chemotherapy to work, patients need the largest possible dose as early as it can be given. Upping a dose by as little as ten percent can raise the chances for cure by as much as fifty percent. The margin of safety for chemotherapy drugs is precipitously narrow, however, so doctors work hard to avoid serious side effects. Using time as a shield can increase chemotherapy's margin of safety, allowing larger doses to be used.

A recent project in France, for example, found that by using time-programmed drug pumps with careful scheduling, patients could handle twice the usual dose. The impact on their tumors increased two- to threefold. Such implantable pumps can deliver as many as four different drugs, while allowing the patient to stay outside the hospital. This saves the expense of several hospitalizations, and it lets patients continue the comforting patterns of everyday life.

Even when a cancer cannot be completely eliminated, because it has spread throughout the body, using timing may extend a life and increase its quality. Many patients with widespread kidney cancer, for example, live fewer than six months after diagnosis. By carefully scheduling chemotherapy, however, one physician recently managed to nearly triple that time. While they could not be cured, these patients lived much longer than anyone expected.

Thus, doctors who treat cancer can now make decisions based both on what cancer they're treating and the best time to treat it. They can take into account annual cycles in the immune system's hardiness and the body's abilities to withstand radiation and drug therapy. For Noreen, and for the one in four of us who will face the terrible choices cancer brings, all these factors can combine to improve the odds of living a long life. And a healthy one.

19

Shadows in the Mind and the Promise of Light

Discoveries about the body's rhythms are changing how medicine views heart disease, cancer and many other physical problems. Meanwhile, new questions press forward about how inner cycles affect our emotional and mental health. Scientists don't yet know a great deal about how biological rhythms interact with mental illness, but proof that they do comes from new frontiers of treatment.

When researchers change psychiatric patients' sleep schedules, for instance, their symptoms diminish. In some cases symptoms entirely disappear. If an illness can be cured by changing the body's schedule, its causes or effects must somehow be tied to cycles playing out over time. The body's rhythms, it appears, also play a part in how we react to everyday stress, in depression, and in our moods at certain times of the day, month and year.

Too Little Time

If the body has a beloved enemy, it is certainly stress. Stress can keep us alive—charging up our energies, allowing us to escape danger or work hard to get food—yet too much stress can be lethal. It lowers our immunities, aggravates heart disease and ages us prematurely.

For many years, chronobiologists overlooked the effect of stress on biological rhythms, and when they first recognized it, their suspicions arose only indirectly. In a few experiments, research subjects were turning in cycles that didn't look quite right. These people appeared to function on excessively long days—perhaps twenty-six or twenty-seven hours long, instead of the usual twenty-four or twenty-five. The experimenters rechecked their results. Everything fit. Then someone remembered the monkeys.

It was unethical to subject humans to severe stresses—painfully loud noise or uncomfortable heat, for example—but this limitation did not hold for animals. When tested under stressful conditions, monkeys often showed disrupted daily rhythms, as if tension had spun the hands of their inner clocks wildly awry. When researchers looked back at the experiments in which humans also displayed unusual cycles, they picked up a common thread.

In some cases the stress was no greater than giving research volunteers an exceptionally large number of tests or awakening them for a few minutes at night to draw their blood. In others, participants had to take charge, making sure that the experiment worked or hooking themselves up to electrodes each time they slept. It quickly became clear that stress can extend our body's idea of how long a day takes.

Under prolonged tension our twenty-five-hour cycles lengthen by two or more hours. This gives us a survival advantage, since we stay alert even after the day ends, but lengthening our days also charges a price. We get less sleep and feel more anxious. Meanwhile, briefer internal cycles can shorten up, making us feel hungry every sixty minutes rather than the usual ninety minutes, or feel tired all day but pop awake in the dead of night.

When we're under pressure, we also face a risk of having our inner rhythms uncouple from one another. The more severe the stress, the greater the likelihood that our sleep cycle, for example, will fall out of tune with our temperature cycle. As the day winds down, our brains are gearing up; when meals are served, our systems aren't ready to digest them. This may account for the insomnia and digestive problems that often come with long-term stress.

No one knows yet if desynchronization of inner rhythms contributes to the frayed nerves of burnout or to the many physical symptoms of tension—headaches, muscle spasms and heart palpitations. We do know that when stress is too severe or continuous, it can sap our emotional strength and light the wick that flares into full depression. Here, again, our inner rhythms play a role.

Darkening Days

Everyone has ups and downs, but clinical depression feels like being caught in an endless "down" with no hope of rising. Clinically depressed people get little pleasure from activities or relationships, cannot concentrate, feel worthless or guilty, and find that their eating and sleeping habits change. Depressions tend to follow their own courses, passing spontaneously after months or years, only to return and settle in again. For some, however, the seemingly endless stream of dark days goes on too long—fifteen percent of all depressed people take their own lives.

Depression's dark days follow cycles so consistent that they are a part of how doctors diagnose this illness. Recently researchers began looking at how these abnormal cycles link up with sleep patterns and chemical changes inside the body.

Frank's case provides a typical example. In his mid-thirties, around the age when first depressions often occur, Frank unexpectedly found himself laid off, and with a new baby on the way. He located another job, but the panic and guilt of those weeks of unemployment never quite lifted.

Frank began waking before dawn, weighted down by guilt that he could not be a more reliable provider. Even the birth of his first child did little to dispel depression's cloud. As months passed, life began to lose all its spice. Food tasted bland, friendships lost their luster, and lovemaking felt like too much trouble. Frank withdrew into awkward, irritable silences, often spending night hours brooding over minor errors he might have made at work or problems his moods were causing at home. Sleep, when it finally came, was restless and fragmentary, a pale shadow of a good night's rest. Meanwhile, inside Frank's body, unusual changes were taking place.

When he settled into bed at night, the electrical waves in his brain took an unusually long time to change to the pattern of sleep. He finally did drift off, but instead of allowing the typical ninety minutes for lazing through sleep's early stages, Frank's brain went straight for stages 3 and 4. He reached REM sleep within half an hour, began dreaming, and experienced more rapid eye movements than normal. This first REM cycle also lasted longer than it should. When it ended, another arrived within twenty minutes, more than an hour ahead of schedule.

These changes meant that Frank would wake up more easily. During the first third of the night, when most people sleep deeply, Frank was quickly cycling in and out of lighter REM sleep. During later hours, he once again experienced an "upside-down" pattern, with far too few REM cycles. In part this may have happened because his temperature and cortisol cycles, too, were far from normal.

While he slept, Frank's temperature did not fall as steeply as it should, so his body stayed warmer all night. The higher the body's temperature, the shorter the time to the first REM cycle, and this may be why Frank began dreaming so soon. In addition, after never having dropped to a normal low, his temperature began rising too early, perhaps rushing him toward alertness again.

Frank's heightened anxiety may also have occurred because cortisol began flowing into his body earlier than it should. Throughout the day and night, Frank's system had more than the normal amount of this hormone, which makes us feel alert. In Frank's case, it made him hyperalert and anxious.

Overall, Frank's inner rhythms followed a fairly average pattern, but they followed it carelessly. Compared to the typical profile, the daily rhythms of depressed people show the same hills and valleys, but their highs and lows look flattened out or squashed. These lower-amplitude rhythms can more easily go the wrong way at the wrong hour, becoming desynchronized. In addition, Frank's temperature and cortisol cycles geared up ahead of schedule, as if in a hurry.

Findings like these have recently sparked new explanations of depression. Some scientists suspect that its rhythms are significantly unstable, perhaps shifting dramatically from day to day. This could bring the feeling of having better days and worse days, but few completely normal ones. The unstable-amplitudes theory may even explain why rates of depression differ between men and women.

In general, men have higher-amplitude daily rhythms that are more stable. Women, until they reach menopause, have less pronounced amplitudes, and premenopausal women do experience depression twice as often as men do. After menopause, women's cycles become more stable, and their rate of depression also declines, approaching that of men.

Another theory suggests that in depression, the body's clock operates too fast. If a depressed person's inner times passes too quickly, he may feel constantly at odds with outer time. This could happen to all the rhythms, or perhaps to only a few—the sleep cycle keeping its normal pace, for instance, while others run ahead of it. Such desynchronization would bring depression's hallmark symptoms of restless nights and leaden days.

To test this theory, one experiment took a group of undepressed people and had them fall asleep later each night. By delaying their sleep, this advanced other cycles in relation to it. Psychological tests of these "normal" people on subsequent days showed them experiencing depressive mood changes. Another study worked with depressed patients, arranging for them to fall asleep six hours earlier than usual. By moving the sleep cycle forward, this might bring it into line with other accelerated cycles. In response to the new timing, these patients' depressions lifted for as long as two weeks.

Promising results are also arriving from a most unusual approach— preventing depressed patients from sleeping at all. For more than half of the people tested this way, skipping even a single night's sleep immediately lifts the depression. Unfortunately, after the next sleep, or even a brief nap, depression returns.

Since total sleep deprivation is hardly a workable cure—we may begin hallucinating after as few as four sleepless nights—scientists are working hard to refine this approach. It now looks as though staying awake for only a portion of the night, the latter half, works as well as skipping sleep altogether. Perhaps this is why, in that experiment that set bedtimes earlier, the patients' moods improved. By going to sleep six hours earlier, they awoke and stayed awake during night's darkest hours.

Much work remains to be done to find out how sleep, temperature and depression interact. We do know that when patients recover—in response to medication, therapy or merely the passage of time—their temperature cycles also become more normal. All these clues suggest that understanding abnormal timing may point toward a cure for the estimated ten million Americans who struggle with depression.

When the Blues Come and Go

Not all depressive moods are chronic, lasting steadily for months or years. For some people, unhappiness has a way of arriving on a predictable schedule—daily, monthly or even annually. These conditions are more brief than chronic depressions, but no less agonizing in their power to provoke sadness.

Clocklike regularity distinguishes three conditions tagged with acronyms: CCO, PMS and SAD. Carbohydrate-craving obesity, premenstrual syndrome, and seasonal affective disorder all appear somehow tied to our hormonal cycles.

Carbohydrate-craving obesity at first looked like an eating disorder, and only later did researchers note its rhythmic profile. People with CCO eat the usual amounts at mealtimes, but then begin snacking on carbohydrates in late afternoon or early evening. For a normal eater, a single cookie may look good around sunset. In CCO a whole box of cookies looks absolutely essential. More than half of all those with CCO are obese, and scientists estimate that as many as two thirds of all obese people have CCO.

Those with CCO say that it's not hunger driving their desires. Nor do their binge foods taste particularly good. Rather, a huge helping of carbohydrates seems to fight off fatigue, anxiety and depression, as if they were abusers of an illegal drug who periodically undergo carbohydrate withdrawal. People with CCO are highly susceptible to depression, and researchers now suspect that their binge patterns may be tied to the timing and levels of chemicals in their brains.

Like melatonin, which makes us sleepy, the hormone which affects our desire for carbohydrates—serotonin—responds to light and darkness. As the day passes, serotonin flows into our bodies and slows our desire for carbohydrates. Also like melatonin, serotonin colors our alertness and mood, inducing feelings of calm and happiness. It contributes to the drowsiness and satiety we experience, for example, after a carbohydrate-rich Thanksgiving meal.

Those with CCO, however, experience a very different effect. After a huge carbohydrate meal, they are refreshed and lively, their depressions lifted. It's as if their natural serotonin cycles were disrupted so that they don't know when to quit.

Drugs that block the effect of serotonin can give people the symptoms of CCO, causing them to crave carbohydrates and gain weight.

Other drugs mimic serotonin's effect, and these help people with CCO skip snacks, allowing them to lose weight. As we understand more about CCO and what makes serotonin slip from its normal schedule, we should learn a great deal about mood, hunger and risk of obesity.

Monthly Lows

Part of this knowledge may also explain why women are much more likely than men to experience cravings for carbohydrates. Women are also susceptible to another cycle of anxiety and depression—premenstrual syndrome. In PMS, too, serotonin's cycle may play a role.

PMS affects ninety percent of all women at some time in their lives, and for ten percent it is severe. Its symptoms of irritability, fatigue and achiness strike predictably, as if PMS were a prescheduled monthly depression. These symptoms last anywhere from a few days to two weeks, then lift when menstruation begins. It is intriguing that one of the distinguishing symptoms of PMS is an increased desire for carbohydrates, and that drugs raising the level of serotonin also alleviate the symptoms of PMS.

In addition, recent research suggests that the cycle of another hormone—melatonin—may contribute to PMS. Women who experience severe monthly lows have less melatonin in their bloodstreams throughout the month. For them, this hormone stops flowing earlier in the night, as if their cycles were running too fast.

When PMS researchers noticed the resemblance between this pattern and the accelerated rhythms of chronic depression, they decided to try a novel approach. Partial sleep deprivation worked for depression; perhaps it would also work for PMS.

They arranged for women with PMS to skip a night's sleep. It was a small experiment, but eight of ten improved. The response was even more pronounced when these women avoided sleep only during the night's latest hours, the same pattern which helped those with chronic depression. It now looks as if light cycles, and the hormones which respond to them, participate in PMS.

No one knows yet whether tinkering with the timing of hormones or sleep can offer cures for CCO of PMS. The possibility does look promising. In addition, researchers are intrigued by the fact the PMS often strikes more severely during the winter. This mirrors a longer, seasonal pattern experienced by millions of Americans—seasonal affective disorder.

The Season of Sorrow

When the short days of winter come, many of us feel sleepier and less energetic. This is a normal decline. For others, the coming of dark days brings profound misery. They have seasonal affective disorder, an annual depression that strikes as many as one in fifteen of us. To see its effects, we can look at the experience of Joan, a divorced mother of three.

During most of the year, Joan feels fine. She works full-time as a bank manager in downtown Minneapolis, and as a single mother she also manages to keep her twelve-year-old and two teens on a reasonable schedule—no small accomplishment. During summers, the family goes boating, hiking and camping. During winter, however, Joan's children know not to expect their mother to join them at a movie or a holiday party. Each year when October rolls around, Joan's mood begins to plummet.

When Joan is in one of her "moods," as her children call them, she has a hard time getting moving in the morning. Minor mix-ups at home—a misplaced T-shirt or broken dish—can provoke her tears. Problems at work loom large, too, and once dinner is served and the dishes done, Joan wants nothing more than to head for bed. Exhaustion is no stranger to single parents, of course, so when Joan first mentioned this problem to her doctor, he advised her to get more rest and rely on the children for chores.

Yet Joan is also experiencing other difficulties. Over the winter months she finds that carbohydrate snacks, second helpings and junk foods look irresistible. She usually gains fifteen to twenty pounds between November and February. Fortunately, the extra weight melts away over summer.

When she's depressed, the only thing that seems to help, oddly enough, is going shopping at a particular supermarket. This store is located inconveniently far from home, and her kids hate going there because it is huge and garishly bright. Yet for some reason, shopping its wide aisles, lit intensely by banks of bare fluorescent bulbs, perks Joan up. She shops there a couple of times a week over the winter, sometimes for only an item or two, but returns to neighborhood stores in spring and summer.

Over the past decade Joan has tried different approaches to beating her bouts with the blues. Sleeping more didn't help, so she went into therapy. Months of counseling failed to lift her mood, so she tried medication. It only made her feel doubly tired. Last year, however, she took a week off in February—the low point of her emotional year—to visit her parents in Florida. Perhaps because of the pleasure of being near family, the depression evaporated. It returned a week after she got home, but only briefly. Then the weather warmed up and she could get outside and be more active again.

Although Joan is unaware of it, she is one of an estimated ten million Americans with seasonal affective disorder. Researchers suspect that another twenty-five million experience some of SAD's symptoms—lethargy, depression, carbohydrate craving and weight gain. While medical reports have hinted at SAD throughout history, this disorder was only officially recognized by the American Psychiatric Association in 1987. Since then it's become increasingly clear that SAD, like CCO and PMS, represents changes in our inner time.

Beginning in autumn or early winter, people with SAD spend increasing numbers of hours sleeping and feel washed-out, listless and uncreative. This SAD-ness gradually lifts in spring, so most fail to see the pattern of their illness, blaming interpersonal problems or their own failings. When they do seek help, people with SAD are often misdiagnosed as neurotic or depressed and treated with a wide range of medications.

SAD is not equally prevalent through the country, as researchers learned when they published a survey in the newspaper *USA Today*. The questionnaire never mentioned SAD by name, but asked questions that would target its symptoms. In replies postmarked in the southern United States, very few people reported symptoms. Replies from more northerly states, like Maryland and New York, showed increasingly higher rates. Further studies have enabled researchers to draw a ''map'' of SAD.

Residents of Florida face less than a two percent risk of developing this winter depression. A thousand miles north, however, in Maryland, the rate more than triples. It reaches eight percent for New Yorkers, but they are fortunate compared to the residents of Fairbanks, Alaska.

There an estimated twenty-five percent of the population experiences at least some SAD symptoms. In Minneapolis, Joan resides in a danger zone, with a SAD risk of perhaps ten percent. Her trip to Florida cheered her up, in part, because she experienced more hours of sunlight each day.

Sunlight appears to be the crucial factor for SAD—the farther north one lives, the greater the chance of experiencing SAD's symptoms. This is so because latitude determines how many hours the sun shines each day, and the length of daylight tells our inner clocks how much melatonin we need.

Individuals vary in how they respond to light, as well as to their own hormones, so it's as if some Minnesotans, and even more Alaskans, were getting melatonin overdoses. A few of Joan's coworkers may feel even more depressed and lethargic than she does. Others may view the arrival of winter only as a great chance to go ice fishing.

Like many conditions tied to our biological cycles, SAD runs in families, with nearly seventy percent of those who have it closely related to someone else with the disorder. SAD is also three times as common in women as in men, although the statistics may only be showing that women are more likely to seek help.

While SAD is easily mistaken for chronic depression, its characteristic rhythms tell a different story. Unlike those who are chronically depressed, people with SAD usually have no difficulty falling asleep. Nevertheless, they experience less slow-wave or deep sleep than normal, giving over the extra hours to REM sleep, or dreaming. When SAD is at its worst, these changes can add two and a half hours to an individual's nighttime rest.

Winter weight gain is another classic symptom, and it can add as many as seventy pounds. Many people with SAD also show lower than normal levels of serotonin, which may account for that compelling desire to eat carbohydrates. By eating extra helpings of pastas, breads and potatoes, they boost their serotonin levels, temporarily improving their moods. SAD patients also respond to drugs that stimulate their bodies' production of serotonin.

It's not yet clear how many inner cycles become altered in SAD, but its annual and geographical patterns confirm one early suspicion. SAD is an exaggerated response to winter's decline in the hours of light. This insight, along with new developments in the technique of using light for therapy, recently led to a cure.

Taking the Light Cure

If lack of light is the problem in SAD, the sensible approach would be to get more. Light definitely works as an antidote, but not just any lighting arrangement will do. The intensity of light, as well as its timing and duration, is critical.

Phototherapists—specialists in light treatment—measure their medicine based on a unit called the lux. Typical outdoor light equals about 10,000 lux, but a bright day can reach more than 80,000. When we step outside on such a day, our eyes adjust, or we may reach for our sunglasses. When we're indoors, our eyes also adjust, preventing us from noticing how dim most interiors truly are. Typical indoor light equals only 500 to 700 lux, a mere twentieth of what we see outside.

Light affects how much melatonin the brain produces, and our brains need to be signaled by more illumination than winter days and wan interiors give. Five hundred lux has no impact, while 1,500 lux works only partially. For phototherapy, patients need to be exposed to at least 2,500 lux, and this requires special fixtures.

A typical phototherapy light box consists of a two-by-four-foot metal case. It holds as many as six fluorescent bulbs. Most phototherapy fixtures use full-spectrum bulbs, which reproduce the wavelengths of sunlight, but plain white fluorescents may work as well. A reflecting surface behind the bulbs amplifies their intensity, and a plastic diffusing screen in front spreads the light evenly. Other designs are U-shaped bulbs, constructed to allow users to tilt the box at convenient angles, and may even be miniaturized—built into caps with small lights attached beneath the visor.

Most patients start out with thirty-minute sessions, sitting about three feet from the light and looking up at it occasionally while they read or work. The length of time needed may increase during November's and December's darkest days, perhaps to as many as six hours. As days lengthen, sessions can be shortened, and when spring arrives, those with SAD can switch off the lights and take advantage of longer natural days.

Sitting in front of lights helps SAD patients feel as energetic and awake as they normally do in spring and summer, but the effect requires four or five days to take hold. Apparently various clocks inside the body need this long to gradually cycle to new settings. Nor is the effect permanent; SAD returns if phototherapy is stopped. In one early experiment a patient became suicidal when treatments were discontinued, but recovered after three days of using the lights again.

Up to eighty-five percent of SAD patients do respond to phototherapy, and for about half of them the therapy works best when it's scheduled in the morning. Since another third get the most relief from midday or evening sessions, individual patients may need to be "chrono-typed" to determine when light will work for them. For this reason, as well as the risk that the depression is unrelated to SAD, professional treatment is far safer than the do-it-yourself approach.

Fortunately, well-managed phototherapy has few side effects. Some people get headaches or eyestrain, others find that their light sessions make them hyperactive and unable to sleep; but reducing the length of sessions usually solves these problems. For most, sitting by the light boxes brings a feeling of calm and alertness. It also appears less expensive and more convenient than flying off to Florida or trolling the aisles of overly bright grocery stores.

The 2,500-Lux Cure-all?

Phototherapy is now widely used for SAD, but this may not be its only application. Numerous other conditions might respond to light, a possibility that prompted the National Institutes of Health to spend more than fifteen million dollars investigating this therapy in 1991. Promising results are beginning to emerge from investigations of the effect of light on everything from immune system problems to binge eating, from withdrawal from alcohol or cocaine to the monthly lows of PMS.

Women who get SAD, for example, are also more likely to experience PMS. Many compare their premenstrual symptoms to those they feel during their winter depressions. Preliminary studies now show that two hours of evening light helps women with PMS.

Light may also help in chronic depression. One experiment exposed patients to evening light and found that their symptoms improved. Eventually the standard treatment for depression may combine antidepressants with light therapy.

Manic-depressive illness also offers a strong candidate for phototherapy. In this illness, episodes of depression and hyperactivity alternate in cycles as short as every forty-eight hours or as long as a year. These cycles parallel changes in temperature and sleep and, like depression, this illness is far more common in women. Both manic-depression and chronic depression may represent a fast temperature cycle getting ahead of a normal sleep cycle. Lithium, the traditional treatment for manic-depression, slows down the body's cycles.

Patients with schizophrenia also have abnormal temperature and sleep cycles, in some cases accelerated by more than an hour each day. Phototherapy may help reset these individuals' inner clocks, as it already does for some Alzheimer's patients. For them, exposure to bright light cuts down on night wandering and "sundowning," a pattern of late afternoon and evening agitation. Someday light treatments may provide an alternative to medication for these people.

Putting Time in Its Proper Light

Slightly more than a century has passed since Edison introduced the tool phototherapists use for their treatments. Less than a decade has gone by since researchers realized that the judicious use of light can alter our emotional cycles. Those time spans are stunningly brief within the history of science—mere pops of a flashbulb.

As we learn more about the impact of light and about what happens when the body's cycles lose touch with it, medical approaches to both body and mind will change.

Change will not stop at the doors of clinics and hospitals, however. Looking into the future, we can see these discoveries affecting nearly every realm of life. From aerospace to architecture, from work shifts to zoo designs, new knowledge about biological rhythms is putting the future in an entirely new light.

V

What May Come

20

Leaving the
Laboratory Behind

Today insights about inner time are moving out of the laboratory
and into daily life. They help infertile couples conceive children,
give shift workers better schedules, and provide therapy for those with
emotional problems. These applications are modest beginnings, however, compared to the changes that knowledge about inner time will
bring in decades to come.

To see what our future might look like, we can fast-forward into the
twenty-first century. There, we may once again imagine the lives of
our average couple from the twentieth century—Norm and Norma.
Back in the 1990s these two lived by the dictates of clocks—setting
alarm clocks, punching in at time clocks, and consulting their wristwatches. In our futuristic scenario, instead of asking the time, people
will typically ask, "About what light would you say it is?"

A New Light On Time

It is a January morning in the twenty-first century. Around 6 A.M.
Norm and Norma—and Son of Every Dog, of course—awaken to the
gradual increase of light in their bedroom. This artificial brightness is

rosy-gold and rises along with the sound of birds chirping. Outside the window it's cold and dark, but the interior of their bedroom suggests a springtime dawn.

Norm and Norma awaken, refreshed after only a few hours' sleep. While Norm showers, a high-intensity fixture above the tub gives him a light-bath. He dresses and takes a few minutes to chat with his wife over breakfast, aware that social contact reinforces the synchronization light signals provide.

When they arrive at work two hours later, Norm and Norma pick up where the night shift left off. Nowadays most companies operate around the clock, but there's no problem with lowered efficiency on the late shifts. Shift workers reset their biological rhythms with light treatments and pills. Both Norm and Norma work the day shift because they are past the cutoff age for night or evening work. Unfortunately, chronobiology never did find a cure for growing older.

Based on a series of preemployment tests, these two schedule their days around their best times. They attend meetings and make phone calls in the morning, then turn to routine tasks during the afternoon. Norma's chrono-profile demands that she nap each day between 3:00 and 3:30 P.M. Concerned about efficiency, her supervisor complains if workers skip their assigned naps, so Norma always takes hers.

Toward the end of the afternoon, Norm leaves his office to visit his doctor. His annual exam has been scheduled at this hour, and on this day, as the best time to check his blood pressure and cholesterol. For the past week he wore a small wrist device, and at his doctor's office this "chronometer" prints out seven days' worth of data on temperature, heart rate, blood pressure and the pattern of the chemicals in Norm's bloodstream. He checks out perfectly, as average as ever. His wife will go for her annual exam in the spring, the best season to catch illnesses linked to her genetic and chronobiological profiles.

Our average couple eats dinner out, at Chez Tres Ordinaire, of course. Norm enjoys spring lamb—tender and fresh despite being, technically, out of season. Artificial lighting schedules have revolutionized animal breeding and agriculture so much that "out of season" is an olden phrase one finds only in books.

Our average couple heads home early because Norm is catching a flight to London tomorrow and wants to begin resynchronizing. In the car he pops the melatonin pill supplied with his airline ticket. Tomorrow he will arrive at Average Products' British Division already on London time.

By early evening Norm and Norma are home. Norm heads for bed, already drowsy. His wife watches the news while taking her twenty-minute DPLT—Depression Preventative Light Treatment, a regime recommended for people of her chrono-type. Before heading for bed, she feeds Son of Every Dog a can of Zeitgeber, "the pet food for the times of your dog's life."

What's on the Way

No pet food called Zeitgeber exists yet, but other innovations in this story have been tried. Amazingly, they all work. A few devices to keep our inner clocks in tune are already on the market. Many more are being tested. Today, as our electrified world continues to accelerate its breakneck pace, inventors, architects, industrial psychologists and medical researchers are finding new ways to help us make the most of inner time.

Our average couple awoke feeling alert on that January morning thanks to their "tropical dawn machine." A researcher working in Anchorage, Alaska, invented this device. To avoid the desynchronization so frequent in northern latitudes, he linked his computer to his bedroom lights. The sun literally rose and set at his preprogrammed command, coming up early in the morning and gradually sinking below the horizon sixteen hours later. He also arranged for bird songs to accompany its arise; he liked the effect of a Caribbean dawn.

Dawn machines can provide the illusion of year-round summer, but other innovations will also change the lighting in our homes. Alongside considerations of floor plans and doorway dimensions, architects can incorporate measures of how light affects our alertness and mood. Illumination throughout the home may be controlled by outdoor sensors that measure the length and intensity of daylight, then automatically adjust light levels in each room. Built-in fixtures for combating SAD may provide selling points for homes built north of a certain latitude.

Nor are individual homes the only places where careful planning can enhance the effects of light. Recently architects, city planners and mental health professionals formed the international Winter Cities

Organization to study the best ways to design cities in the far north. Canadian and U.S. researchers have joined with scientists from the former Soviet Union to explore how residents of polar regions can use light to counteract the effects of extended winter darkness.

Some ideas to achieve this are already in use. When Norm took his morning shower, an overhead fixture, based on an idea tried out in Europe, gave him a "light-bath." By exposing coal miners to periods of light, authorities found that they could lower rates of illness and absenteeism. Norm's conversation with his wife over breakfast was also part of a routine that specialists call "chronohygiene." In the future we will become more aware of the effects of such time givers as social contact and meal timing, and we will probably use them to keep our bodies and minds working at peak efficiency.

Chronohygiene may be implemented first for the elderly. One research project in a veteran's home near Boston revealed that living under full spectrum lighting helped residents' bodies absorb more calcium from their meals. Other investigations have found that natural light also strengthens muscles. In retirement homes, senior centers and assisted-living programs, phototherapy will probably also be used to help older people avoid the sleep disturbances that are so common late in life.

Doing Business Around the Clock

At the office Norm and Norma could relieve the night shift because those workers used phototherapy and medications to adjust their bodies to shift work. The technology to do this is already available, but ethical questions affect the decision to use it.

By avoiding light during nonwork hours, evening- and night-shift workers can already fool their inner clocks into believing that they are living on normal schedules. Workplace cafeterias can offer certain foods at the right hours, maintaining higher levels of alertness in employees. Shift workers can put on dark glasses when they head home and can stay on their work schedules on their days off. In the future, evening and night workers may also put their home-lighting timers on nonstandard schedules, work under banks of intense fluorescents and perhaps take regular doses of melatonin or other drugs to maintain their unnatural routines.

Workers on rotating shifts will be especially good candidates for phototherapeutic or pharmaceutical resetting, and those who want to advance their careers could face powerful pressure to reset their bodies. Meanwhile, chrono-typing, already available by means of simple psychological tests, may play a role in preemployment testing.

Chrono-typing reveals whether a job applicant is a day or night person and can also show the amplitude of an individual's rhythms. Since lower amplitudes allow cycles to fall out of line more easily, employers can screen out candidates who are easily desynchronized by schedule changes. If put on shifts, these people would be less efficient and more prone to cause accidents. Chrono-typing looks like a boon to employers, but selecting workers in this way also presents serious legal risks.

Chrono-typing can stigmatize older workers and mark those with low-amplitude rhythms as less than ideal employees. With each passing birthday, many who could previously adapt to night work become less flexible in their rhythms, and this issue may become particularly intense as the baby boom generation ages. Chrono-typing can make some of us, as one future-watcher put it, "chronobiologically unemployable," confronting society with questions about discrimination based on age or inborn rhythms.

For those who work in factories or provide twenty-four-hour service in hotels, television stations or telephone companies, such discrimination may seem unjustified. The question gets trickier when we consider the dangers of desynchronization among airline pilots, health workers, and nuclear power plant employees. We permit mandatory testing for drugs and alcohol. It may soon be acceptable to monitor certain workers for the strength of their rhythms.

A few moments on an EEG can reveal whether employees have followed their assigned sleep schedules. If police officers or nurses must rotate shifts, should they be required to resynchronize within a set time span? If an inexpensive skin patch could reveal body temperature, should off-cycle workers be sent home, and should their paychecks be docked?

These are difficult questions, and we can only hope that answers will emerge before many Norms and Normas live in the twenty-first century. Already some workers are overcoming the risks of being off cycle by the use of "polyphasic sleep."

Emergency crews responding to disasters such as earthquakes and hurricanes are sometimes put on twenty-four-hour work shifts. Polyphasic sleep schedules assign them to take short naps at times when their performance levels will be lowest. Because these sleep-deprived workers doze off at the right times, their naps provide primarily REM sleep, enabling them to awaken refreshed and alert, ready to make logical, life-saving decisions.

Checkup Time for Medicine

On that twenty-first century day when Norm went for his physical, his doctor would know more than we do today about biological rhythms. Yet before that can happen, more than technological innovations are needed. It's taking a long time for the latest discoveries about our inner time to change what happens in doctors' offices, because attitudes need to change too.

Pioneers in the study of chronobiology see it as the latest of the so-called integrating sciences, comparable to evolution, developmental biology and genetics. They believe that knowledge about inner time will contribute as much to medicine as did the practice of scrubbing before surgery or the invention of the high-powered microscope. Yet progress in applying their discoveries is slow, because our health care system is highly specialized. It is also tied to traditional scheduling.

A cardiologist may know, for example, that most heart attacks occur in the morning. Yet he may not recognize the role that light plays in synchronizing the body. Specialists in another department, the sleep lab, know about light. Meanwhile, the nutritionist, relegated to an even more distant professional zip code, knows that biological rhythms can aggravate a weight problem, putting extra strain on the patient's heart.

In an age of specialization, chronobiology sees the body as a whole, integrating knowledge from anatomy, biochemistry, microbiology, pharmacology and other fields. This puts greater demands on physicians seeking to integrate chronobiological findings, and most medical schools give only a minimum of time to teaching the central concepts of inner timing. Around the world, several dozen centers are now dedicated to chronobiological research, but only about a dozen of them are in the United States. Some European countries have been developing this new field more quickly than the United States has, a trend that disturbs many American researchers.

In addition, in the United States and elsewhere, potentially life-saving discoveries face challenges from medicine's traditional scheduling. A clinic that's open from nine to five is unlikely to detect high blood pressure in the patient who experiences it only at night. The hospital radiation unit that must fit dozens of cancer patients' treatments into an eight-hour day cannot easily treat a tumor at the hour when its cells are dividing fastest. Even the pharmacist, typing instructions for pills to be taken at certain times, may neglect to inquire whether this patient works nights or frequently travels across time zones.

Despite these obstacles, new information about our bodies' schedules will gradually change how traditional medicine is practiced. Many physicians already take advantage of the greater precision that timed approaches offer, and in the future even more will do so. Diagnostic tests become less expensive when doctors know the best times to get significant results. Bothersome variations in rhythms become allies that explain illness. As physicians recognize this, they will routinely enlist the workings of our biological clocks to improve our care.

Timing as the Best Medicine

The wrist device that Norm wore to monitor inner rhythms is not yet available, but similar tools are in use. Portable monitors for heart rate and blood pressure help in diagnosing arrhythmias and other heart problems. At-home tools to detect the temperature changes that signal breast cancer are on the market, and patients with asthma, arthritis and chronic pain can use self-tests to chart the rhythms of their conditions.

Among heart attack patients, around-the-clock monitoring has already been used to predict which individuals would die within five years. When all heart patients are checked this closely, those at higher risk can receive special follow-up. Similar devices are also available to monitor gastric activity, temperature and breathing.

Within the next decade, according to one expert, all newborns will be watched during their first forty-eight hours to detect their risk of high blood pressure. Follow-up testing at midlife can observe rhythms, comparing them to established profiles, and anticipate problems. Preventative checkups, such as Pap smears, can be scheduled at times keyed to seasonal patterns of diagnosis, and those who study epidemics can take both geographical patterns of transmission and time cycles into account.

Essential information will also come from looking at several cycles together. Scientists at the University of Minnesota have developed a program that enables the IBM personal computer to interpret changes in a variety of rhythms. Computer software now in the works will take the data from around-the-clock readings and compare it to average profiles. When home computers can use similar programs, we may be able to detect many illnesses even before symptoms announce their presence.

By Norm and Norma's twenty-first-century day, the drugs they take will come with chronotherapeutic package inserts, individually tailored to the patient's inner rhythms. Those with chronic conditions, such as diabetes or angina, can use portable devices or have drug pumps implanted to keep a time-specific level of medication in their bodies. Implantable devices may eventually link up to "closed-loop systems," measuring blood chemicals or hormones, then self-correcting the schedule based on the patient's condition. Those with ulcers, for example, may automatically receive medication when their levels of gastric acid are highest and at specific times of day.

Eventually the design of hospitals will also include chronobiological considerations. New monitoring tools can release staff from tedious tasks, and lighting schedules may become part of physicians' orders. One experiment, for example, compared an intensive care unit without natural lighting to another whose windows let in light from the outside world. Patients in the windowless room more frequently lost track of time and remembered their stay less well. They also had more trouble with sleep disturbances and more hallucinations. Someday not only intensive care units but hospital nurseries may use routines of light and darkness to help recovering patients and newborn babies adapt before going home.

Transplanting Time

These chronobiological innovations will manipulate light to adjust our inner time. That's an indirect approach, but someday we'll take the next step, using hands-on techniques to reset inner clocks. We already know that a portion of the hypothalamus, the SCN, serves as a master clock for the body, receiving light from the eyes and synchronizing many functions. Researchers are working hard to locate other clocks, including those in individual organs. As new knowledge arrives, so will the temptation to tinker.

The SCN looks like the first candidate for such tinkering. A reading taken from a single cell or one chemical may eventually tell the hour our inner clocks show. When that happens, information about inner time will be gathered as routinely as temperature is taken today. Techniques to reset it may become equally common.

We may some day use an antimelatonin, or "chemical light" pill to reset ourselves. Its first use will probably be linked to the scheduling of medications. A "reset" pill could bring several rhythms into line, so that a therapeutic drug could work inside an ideally timed system.

Those who are easily desynchronized or who operate on days that are too long or too short may want to use electrical or chemical signals as pacemakers. Teenagers and older people, whose sleep cycles are easily upset, may choose drugs or light treatments to stay in tune with the rest of society. Eventually phototherapists and sleep specialists will probably work alongside psychiatrists to help correct disordered cycles in depression and other mental illnesses.

Even those who live on normal schedules may leap at the chance to reset their inner clocks. After all, resetting will offer a cure for that most widespread of modern maladies—lack of time. Many would gladly drop a dose of "chemical time" if it would turn them into one of those lucky people whose inner clocks make them need only a few hours' sleep each night.

A few high achievers might even opt for surgery. Researchers have already altered animals' inner cycles by transplanting SCN's between short-day and long-day hamsters. Decades from now an SCN tune-up or transplant may seem no more remarkable than today's face-lift or tummy-tuck.

Like Norm and Norma, of course, most of us will probably pass up the chance to undergo time-saving surgeries. We'll know more about our inner time, but as long as it's running on schedule, we'll take it for granted. We'll leave our doctors' offices feeling chipper, assured that other technologies have made provision for our dinners at Chez Tres Ordinaire. We will check the light and trust that evolution—or technology's answers to it—are keeping us in time.

Every Meal in Our Own Good Time

Nowadays, when we shop for groceries or eat out, we generally expect certain foods to be available at set times of year. Just as tropical flowers come from tropical climates, winter fruit must be shipped in from the south. Vegetables and meats are better or worse, more or less expensive, depending on the month. In the future, however, new technologies will change what we can find at the market or on the menu.

Careful scheduling of lighting already makes chickens lay more eggs, calves fatten faster and crops shoot toward artificial suns. Lighting regimes can make animals' winter coats thicken faster and jump-start seedlings in nurseries, providing exotic blooms at latitudes evolution never intended them to see. As for Norm's dinner of spring lamb, served fresh in January, if sheep receive melatonin at the right hour, in the right season, their bodies get ready early for breeding. This can increase the number of lambs each ewe delivers.

Biological timing can also help protect plants and animals from pests and diseases. Many parasites have cycles that make them easier to defeat—the common house fly, for example, goes into an instant tailspin if exposed to poison at 4:00 P.M. As for the animals and plants we wish to save, knowledge about inner time may eventually help us preserve endangered species.

The first clue to this possibility arrived, like so many scientific discoveries, completely by accident. Some years ago a zoo in Syracuse was having trouble with nighttime vandalism, and someone suggested installing very bright lights and leaving them on all night. The technique worked, and it brought a result no one expected.

Geese began laying eggs. The deer population skyrocketed. Sheep delivered lambs, and the bear, wallaby and chimps got pregnant. The cougars delivered their fourth litter. As one worker put it, the place turned into a veritable "maternity ward." Fooled by the apparent length of daylight, the animals thought spring had arrived and behaved accordingly.

On that twenty-first-century day when Norm travels to London, he may take a few hours to visit the London Species Preservation Park. In this natural setting—which replaced the London Zoo in the late twentieth century—he will see snow leopard cubs, newborn white-tailed deer and eaglets, all bred through interactions between hormones and light.

Times to Come

As we look even further into the future, we can travel with Norm across space and time. His trip to London will be quicker and more comfortable, and may even include better airline food. It will certainly provide accommodations to meet the needs of his inner time.

Moving at twenty-first-century speeds, we might imagine his flight taking only two hours. When he arrives, his body clocks tell the same time as Big Ben, thanks in part to that melatonin pill. In addition, shortly before landing, the airline crew provided light goggles to Norm and the other passengers. After they wore these for thirty minutes, the settings of their temperature and sleep cycles could not be distinguished from those of the ground crew at Heathrow.

In the cockpit the pilot and copilot followed quite a different regime. While in flight, they obeyed assigned schedules for polyphasic sleep. Catnapping in the cockpit, which FAA rules used to forbid, put them at peak alertness for the flight's most crucial moments—takeoffs and landings.

In addition, after they leave London, this crew will fly on to Bangkok. London to Bangkok is a short hop these days, but their bodies need to absorb a twelve-hour time shift. Instead of moving their cycles backward, as the passengers did, crew members take light treatments and pills to shift the hands of their inner clocks forward. This way they pick up eleven hours, the twelfth hour coming from the "spare" our twenty-five-hour clocks provide.

This routine for flight crews came from research done on those who traveled even greater distances in the late twentieth century. When nations began putting astronauts into orbit, they had to find solutions to new chronobiological challenges.

For an astronaut in orbit, a "day"—a span of sunlight followed by darkness— may pass in less than two hours. On the first space flights, back in the twentieth century, mission control tried keeping astronauts on ground time. Crew members slept poorly and may have been less than alert if an emergency arose. In addition, to take advantage of the earth's position, launch times often were scheduled in the dead of night. At such hours both astronauts and ground workers were at their least alert.

In the early 1990s the National Aeronautics and Space Administration changed all this. A light chamber set up at the Johnson Space Center used light bulbs and highly reflective paper to expose astronauts to sessions of 10,000 lux, twenty times normal indoor light. By spending time in this room over several days before the flight, crew members reset their inner clocks. With their bodies fooled, shuttle crews found they could turn in peak performances even if ground controllers were yawning away the wee hours.

Looking another decade into the future, we might assume that NASA began resetting not only astronauts' inner clocks but those of ground crews as well. Before a launch the entire mission control team shifted to a peak performance schedule, increasing the margin of safety. As the twenty-first century progressed, fully operational space installations were designed and put into orbit. By the use of lighting, hormone dosing and—for extremely demanding missions—transplanted SCNs, space station workers' clocks were resynchronized on demand.

Epilogue: Once Upon a Future Time

But such a scenario comes long after Norm and Norma's lifetimes. It is perhaps the era of their children, or their children's children. In addition, we may have traveled too far and too fast.

When we left earth in the late twentieth century, we carried reminders that we could not escape the character of our own evolution. Like some chemical essential to life, future generations will carry earth-time outward with them toward the stars. It will travel along inside their cells, their hearts, their brains and their bloodstreams. Yet once we leave our home planet, the old-fashioned concept of days, months and years also falls away at our backs.

On the moon a single day takes a month to pass. Outer space knows nothing of twenty-four-hour cycles, moon-months or the yearly cycle of seasons. The first astronauts' bodies carried that knowledge into the universe, and when permanent space stations require humans to live without earth's reminders of inner time, technology will have to make up the difference.

Scientists are already testing what happens to plants' and animals' cycles when they go into space. Humans living in space will probably use zeitgebers of light, meals, social scheduling and medications to keep their bodies in tune. Eventually we will understand inner time so completely that we will probably reach for tools to abolish it.

The first employees living permanently in space will be true migrants. If market forces have their way, these employees' lives may be as difficult as those of migrant workers today. The immense cost of maintaining space factories and laboratories may motivate employers to find ways to override the evolutionary artifact of natural rhythms.

Workers' inner time could be reengineered so that they need very few hours of sleep, or perhaps none at all. If the idea catches hold back on earth, old habits of living in sync with days and seasons may pass. We would see the practice of living in tune with earth's cycles as benighted and confused, the way we now look at the ancients' celebrations of solstices.

Brave New Otherworld

In such a future, we might imagine the offspring of our twenty-first-century couple. Norm and Norma's genes will have been carried light-years from earth. We'll call this new couple Nrm and Nrma, since by then everyone will be in too much of a hurry to pronounce extra vowels.

As space station employees, Nrm and Nrma work continuously until a job is done, then take time off. There are no clocks keeping track of their work; such things were left behind long ago. In part, their duties depend on solar winds, and since these tend to come and go, Nrm and Nrma occasionally get an unscheduled holiday.

On one such day they decide to clean out the attic of their space station. Squeezed way back in a corner, where they never look and hardly dare to go for fear of those genetically engineered spiders that got loose long ago, they find a battered box filled with old junk. They pull it down and hear something inside it making a strange noise.

Nrm will push aside the old calendars. He will remove the fast-food take-out menus, the space shuttle schedules, and the broken parts from time-saving gadgets. He'll burrow through papers advertising all the time-compressing discoveries of the twentieth century: automobiles and computers and fax machines and bar code scanners and automatic teller machines and slicer-dicers. At last he'll uncover a round object, rather like a dial with two pointers on its face. It has straps on either side, and it is ticking.

Nrm holds it up. Delighted that he knows what it is, he explains that he's read about such objects. People once believed that events had to happen in a certain order, Nrm explains. Objects like this helped society keep in step, and people wore them all the time, even while they slept. They kept extras around the home and the workplace. Fancier, bigger versions of these things might blaze from the tops of buildings, long ago, in the cities of Earth.

Nrma listens, perplexed. Since she and Nrm grew up in an age after chemical treatments and surgeries became standard procedures to override inner time, she has no idea what he is talking about.

It all had to do with sequences, Nrm explains. People used such objects constantly back when there was such a thing as time.

Nrna stares at him, her gaze astonished. By our twentieth-century calculations, a second or two passes before she asks, "Tme? Wht's tht?"

INDEX

185

.